Yale Agrarian Studies Series
JAMES C. SCOTT, SERIES EDITOR

The Agrarian Studies Series at Yale University Press seeks to publish outstanding and original interdisciplinary work on agriculture and rural society—for any period, in any location. Works of daring that question existing paradigms and fill abstract categories with the lived experience of rural people are especially encouraged.

—JAMES C. SCOTT, *Series Editor*

James C. Scott, *Seeing Like a State: How Certain Schemes to Improve the Human Condition Have Failed*

Steve Striffler, *Chicken: The Dangerous Transformation of America's Favorite Food*

James C. Scott, *The Art of Not Being Governed: An Anarchist History of Upland Southeast Asia*

Edwin C. Hagenstein, Sara M. Gregg, and Brian Donahue, eds., *American Georgics: Writings on Farming, Culture, and the Land*

Timothy Pachirat, *Every Twelve Seconds: Industrialized Slaughter and the Politics of Sight*

Kuntala Lahiri-Dutt and Gopa Samanta, *Dancing with the River: People and Life on the Chars of South Asia*

Alon Tal, *All the Trees of the Forest: Israel's Woodlands from the Bible to the Present*

Graeme Auld, *Constructing Private Governance: The Rise and Evolution of Forest, Coffee, and Fisheries Certification*

Jess Gilbert, *Planning Democracy: Agrarian Intellectuals and the Intended New Deal*

Jessica Barnes and Michael R. Dove, eds., *Climate Cultures: Anthropological Perspectives on Climate Change*

Shafqat Hussain, *Remoteness and Modernity: Transformation and Continuity in Northern Pakistan*

Edward Dallam Melillo, *Strangers on Familiar Soil: Rediscovering the Chile-California Connection*

Devra I. Jarvis, Toby Hodgkin, Anthony H. D. Brown, John Tuxill, Isabel López Noriega, Melinda Smale, and Bhuwon Sthapit, *Crop Genetic Diversity in the Field and on the Farm: Principles and Applications in Research Practices*

Nancy J. Jacobs, *Birders of Africa: History of a Network*

Catherine A. Corson, *Corridors of Power: The Politics of U.S. Environmental Aid to Madagascar*

Kathryn M. de Luna, *Collecting Food, Cultivating People: Subsistence and Society in Central Africa through the Seventeenth Century*

Carl Death, *The Green State in Africa*

James C. Scott, *Against the Grain: A Deep History of the First Civilizations*

Loka Ashwood, *For-Profit Democracy: Why the Government Is Losing the Trust of Rural America*

Jonah Steinberg, *A Garland of Bones: Child Runaways in India*

For a complete list of titles in the Yale Agrarian Studies Series, visit yalebooks .com/agrarian.

A Garland of Bones

Child Runaways in India

Jonah Steinberg

Yale

UNIVERSITY PRESS

New Haven & London

Published with assistance from the foundation established in memory of Amasa Stone Mather of the Class of 1907, Yale College.

Yale University Press books may be purchased in quantity for educational, business, or promotional use. For information, please e-mail sales.press@yale.edu (U.S. office) or sales@yaleup.co.uk (U.K. office).

Set in Bulmer type by Tseng Information Systems, Inc.
Printed in the United States of America.

Library of Congress Control Number: 2018940488
ISBN 978-0-300-22280-7 (hardcover : alk. paper)

A catalogue record for this book is available from the British Library.

This paper meets the requirements of ANSI/NISO z39.48-1992 (Permanence of Paper).

10 9 8 7 6 5 4 3 2 1

Contents

Preface

THIS IS A BOOK ABOUT LIFE AND DEATH, about the way that
some of the most vulnerable, least powerful people in the world live history,
the way the forces of empire, the movements of states, the conquest of lands
are imprinted on their very bodies and in the most intimate reaches of affect
and intellect. This is a book about children's journeys, usually quite alone,
from villages to distant cities, about the villages they leave and the cities
they reach, and about their transformation from rural subjects to something
called "street children" along the way. It is about the individual children but
also about their numbers, en masse, why there are so many of them collectively doing the same thing at the same time. It is about the stories they tell
of themselves and their perception by others. In the runaway's thousand-
mile journey I hope to discern, and illuminate, the ways that colonial reconfigurations of self and status, land and power, rule and debt converge on
the lives of those most powerless in global architectures of governance and
production, those who lack even the possibility for citizenship and access
to the weakest forms of legitimate labor. They have access to the machinery
of state for punishment but rarely protection, and unprotected even by kin
they die daily or live in misery. Their pains and passages capture, recapitu-
late, in experience, in emotion and life course, rule and conquest itself,
the vicissitudes and vacillations of polity and capital, the machinations of
modernity's engineers, the failure of bloody utopias, the remainders of rupture. In their vulnerability and subjection, child runaways reveal a remarkable embodiment—and subversion—of these phantom forms.

Acknowledgments

THE FIRST THANK YOU IS DUE TO all the children and former children that this book is about, who participated in the research that constitutes it, and who inhabit, haunt, or speak through its pages. Without them and their remarkable vivacity, this project would never have come to be, and India and the world would not be the same sort of place.

The research for this book was made possible by National Science Foundation grant NSF-BCS-0924506. Profound thanks are due to Deborah Winslow, program director at the NSF's Cultural Anthropology program, who patiently guided me—through more than one failed and ill-conceived application—toward the ideas that would eventually give this research its shape. Once I was doing the research, Deb braved endless e-mails from the field, just as she does now, and answered them generously. Thank you also to Jeffrey Mantz for his support.

The most profound debt of gratitude here is due to Khushboo Jain, without whom the research for this book could and would not exist. Khushboo's efforts, and voice, and character, appear throughout this book. The ethnographic fieldwork she carried out for this project was beyond expectation and compare. She deserves credit for so much in these pages, and also for her dedication to and legal advocacy for the children who populate these stories.

Thanks are due to Beth Mintz, who said, when teaching on my former specialties was all covered, "What else are you interested in?" and then "Why not teach on street children, then?" She was the one to whom I said, "Okay, why not?" Also to Ellie Miller, who thought then and for that reason that my research was already on street children and nominated me for the fellowship whose (failed) application would lead me circuitously to this book. I am indebted also to Sanjoy Roy, a tremendous human being and one of my first mentors for this work in India, as early as 2007.

I did not, I said.

You know Pabna? he persisted.

Neither that, I replied.

Nabil, it turned out, had traveled a solid thousand miles from Pakshi Village, where the bridge spanned the Ganga (there the Padma).[1] The Pakshi of today is a productive agricultural town, with its own Export Processing Zone (at Ishurdi), a Sugarcane Research Institute, a major paper mill, and facilities for railway employees. It is directly adjacent to the site for the Ruppur Atomic Energy Plant. Despite all this industrial activity, it remains distinctly rural. In the longer course of historical event and rural change, that he had come from that particular village beneath that particular bridge is not inconsequential: Pakshi is adjacent to Ishurdi, a depot town, built to serve the imperatives of the imperial British railways. The bridge at Pakshi is the Hardinge Bridge, named for Lord Hardinge of Kent, viceroy of British India (1910–1916). Thus Nabil entered the world, grew up, suffered, and (taking things into his own hands at last) left a place structured, even generated, by the filament of empire—a filament meant to connect rural and urban and still doing that. Indeed, the filament still connects this hinterland to its imperial metropole despite the borders that have been etched into the land, despite even the destruction of the bridge in the wars over that etching (1971, wherein the Hardinge Bridge was bombed).

For Nabil, born beneath this monumental artifact of empire into a family of laborers he described as "middling, neither poor nor rich," circumstances began to change some five years before I met him, when his father's kidneys showed signs of failing. An operation might save him, the doctors said, a procedure, a donation of organs: but such options are out of reach for such a family in such a place, and Nabil's father was left with no choice but to die. And so he did. Women left thus alone are rendered in rural widowhood deeply vulnerable, and in part for this reason Nabil's mother remarried. But Nabil got on very poorly indeed with his new stepfather, and as friction grew at home Nabil took charge (of himself, not the house), packed up, gathered his scant savings from various jobs, and left.

How? I inquired.

Foot, bus, and train, he replied. I got myself to the border.

Was anyone else there?

Yes, he explained, lots.

Any children?

A few boys waited there with me at night, and then we paid and were let across.

Nabil arrived in Delhi, at length, and took up residence in New Delhi Railway Station, an endpoint for many such itineraries. Like the many others who have completed the passage to this point, he began to live on the collection of others' railway-journey waste, particularly plastic bottles. And thus began his life in the city. He had joined the ranks of the children calling themselves *kangaal,* the "bone children," from a Sanskrit term, one rich in religious association, denoting a skeleton or "a garland of bones." There is indeed something darkly and uncannily poignant in the etymology of this term that gives this book its title, in particular for Nabil but in general for the runaway child suddenly alone on the city's streets.

But these are not straightforward journeys, nor simple. Indeed, often they are imbued with sadness and the loss of something or of many things. I could see this on Nabil's body. Taking my cue from years of exposure to very frank body talk, I decided to ask: what happened to your arm?

It got run over by a train, he said. It started up and cut off my arm.

He said this matter-of-factly, without melodrama or flourish, without even wistful regret. Thus the very vehicle (in both senses) that conjured his hometown and its spatiality, the very vehicle that enabled his journey to the postcolonial capital, had marked itself forever on his body: history, in some sense, sliced off Nabil's arm.

I bled and fell unconscious, he continued. A friend brought me to the hospital and saved my life.

Does it make you sad, your arm?

What's the profit in being sad? he replied. What can I do about it now? I survived.

The measure of what the lost arm signified, however, lay elsewhere, in the village, in the home, in the natal space. The severing of the arm was indeed a severing of a certain umbilicus, blocking off forever certain possibilities and conditions of return, even if physical return was to be accomplished, which eventually it was.

Do you want to go home? I asked.

I want to, but—

But what?

I can't.

Why not?

My arm. I am afraid of Mother seeing me, how sad she'll be. I can't have her see me.

And so across the passage of months and years Nabil and his companions spent their days by the Hanuman Temple and the Yamuna Market, near the Birth and Death Registration Office, the old Box Market, and, at its desolate end, a river of sewage flowing through a tunnel formed by a bridge on whose keystone was a faded and forgotten inscription reading

Calcutta Gate
1852

Another bridge, another ghost of empire. Of this spot, Sir Gordon Risley Hearn wrote in 1906 (in *The Seven Cities of Delhi*) that the "Calcutta Gate, built in 1852, has been removed to admit the railway, and this portion of the Trunk Road has been abandoned" (7).

Likewise now. On top of the Calcutta Gate bridge, more than a century and a half later, still lie the railway tracks bearing trains bearing runaways into Old Delhi Depot. I want to suggest that the presence of—and connection between—those two bridges in Nabil's disparate living spaces signifies more than it might at first seem; it is more than superficial, more than a quaint literary coincidence. Its poignancy lies in more than what it intimates as metaphor. In chapter 5 I will dig into that. On these tracks likewise roll trains bound back toward Pabna and Nabil's natal home, starting their long ride, in the opposite direction, to Howrah, sometimes even bearing weary runaways home.

Nabil and his friend Salauddin in those days slept beneath another bridge, the so-called ISBT flyover (an underpass), in the company of a young prostitute and her young family, a number of addicts, and a polio-stricken adult, Masud, who had himself run away as a child twelve years earlier from West Bengal's Sundarbans (in the 24 Parganas division); though the latter had returned home, at length, to "end his family's poverty," as he put it, he continued to live where he had arrived over a decade ago. Nabil's "bed" was beside a great square trough through which flowed another torrential current of mostly sewage, driven by the force of Yamuna River water.

Nabil had not been in touch with his mother since leaving. In fact, she had no word that he was even still in this world. I suggested she might rather see him one-armed and alive than carry on imagining him dead, in uncertainty. A few days later, he disappeared. On the street, in those days, it was nearly impossible to find Nabil by using his real name.[2] He was, rather, known only as Toonda, a term denoting "cripple" or even "idiot," in the archaic sense, and thus his embodied marker, as is common, became a synecdoche of his entire persona. I looked everywhere for him, asking the many scattered small populations of children where he might have gone, and indeed surprisingly many across the city knew him—by this same marker, among other things—but not where he had gone. After my departure, Khushboo, my research partner, continued the search, asking around everywhere for him. Had he followed other children to the cool air of the Himalaya? Had he moved to a different site?

A year later, he was back gathering bottles in Old Delhi Station, two more thousand-mile journeys under his belt.[3] He had gone home to his mother and left again, again choosing the city over that distant village, the bridge at Calcutta Gate over the Hardinge. Someday perhaps I will learn what came between these two moments in that next chapter of Nabil's tale, the murky and anxious space between home and the world. But for now I tell Nabil's story because it illuminates the unique complexity of the runaway's journey, each one unlike the other and yet all of them connected, and that, in turn, illuminates the complex, troubled, and well-traveled space, filled with its hundred million transits, between village and city. Nabil leaves a rural borderland marked by and imbricated with empire for an urban center and finds it densely populated with compatriot souls unhitched from their own sites of agrarian collapse and colonial aftermath. They are not the same places, but they are not dissimilar. Nabil's entire circuit is punctuated by the stuff of history, waypoints, and traces, at every step. Are such journeys the products of a sort of historical predestination, rendered nearly inevitable by the political cartographies of debt, track, and dispossession in which they are situated, or are they the mounting of a form of defiance? If the latter, is the defiance collective, and if yes, is it so perceived, or does it always seem a lonely, solitary passage, isolated from all others?

Runaway Republic: Children's Journeys as History's Incarnations

The question I wish to ask is not only what makes a child leave home, idiosyncratically speaking, but what makes children in certain situations *likely* to leave home, and, further, what are those situations and configurations that tend to produce this likely departure, and why and how widely are they shared. I wish to make sense of whether it is culture, tradition, or rite, on the one hand, that generates child departure or, alternatively, the political-economic forces embedded in history, debt, and rule—or both; I allow, of course, or more than allow, that culture and political economy need not be and indeed rarely are mutually exclusive, and that changing historical conditions may or must produce emergent cultural forms, so a cultural model for departure from home may be a collective response to structural conditions, to social change, circumstance, and crisis. The question that then arises is whether such a response is mounted in the absence or the knowledge of the other sites of its articulation within larger scales or fields of time and space: does a runaway feeling the pressures of his particular situation see himself as joining a highway where he will meet many such others, and where many such others have also already trod, preparing the path for him? If so, does he imagine these others rooted in radically different cultural settings? Does the parent whose household is suddenly missing this child, on the morning that he is not in his blankets or the evening that he never returns from the cricket pitch, think *this is a thing that happens to people like me in places like this?* Or does that parent just feel the texture of the situation: the emptiness of the household, the echoes of the night before, the fears over a child's fate? Regret ferments quickly, and these are fraught and painful moments that cannot always readily be reduced to a formulaic calculus of variables; they are composed of stories, stories the people in them feel are their own. And yet the shared histories are the root and fundament, the necessary antecedents, of these stories.

A number of well-named and expectable culprits are implicated in the production of what we might construe to be a common set of historical circumstances and experiences—largely connected to underlying modes of structural subjugation—surrounding the runaway's departure. The manipulations of identity (à la Dirks 2001), territory, and power under empire, and in particular colonial reconfigurations of rural credit relations, are foremost among them. The imperial district-by-district gazetteers at the British Library suggest a correlation between old colonial debt servitude and the geographical distribution of child departure, but they do not articulate causes. Districts with the most exploitative policies toward poor farmers, even nearly two centuries ago, are among those most likely to see children leave. But a panoramic and careful look at the villages and districts that feed the populations of children on city streets reveals potential other candidate triggers of social crisis: the rescripting of place under the auspices of nation, the reconfiguration of land and status in the rearrangements of the "Green Revolution," the vicissitudes of "development" (see Bardhan 1984, 2003), dislocation for macrostructural rearrangements (as in Baviskar 2003), and regional contestations of power such as in the Naxalite conflict (see Ray 2012; M. Roy 2010; Chitralekha 2012). While varied, these all point to potential catalysts of what I identify as intimate responses to the movement of political economies.

And thus I am of the mind, to return to the essential point, that people live history, that macrocosmic structures are experienced by individuals, but if that is the why then the how still needs an articulation. It is my supposition, after looking long at the patterns, that historical forms translate quite easily, if orthogonally, into *domestic* stressors (or, alternatively, in other cases, into privileges and pleasures) and that a domestic crisis, including abuse, an emotionally violent divorce, or a family's susceptibility to disability and death (see Butler 2010 to contrast Agamben 1998), may well emerge out of widely shared experiences of landlessness, dispossession, administrative manipulations, and violent displacement.

But if running away is to be taken, at least at the surface level, as an

existing cultural form that children may mobilize or manifest, a blueprint
that they follow, then a secondary question becomes *how* does information — instructions, methods, practices, itineraries — get transmitted peer to
peer or (as I will address) through public texts and images? This is a question that Urban (2001) and Agha (2007) have spent much time considering
in attempts to theorize cultural circulation. It seems that rural train depots,
as I underscore in the next chapter, serve as an essential social space for
the transfer and sharing of information about running away, perhaps more
than villages themselves. Indeed, many children who leave have histories of
informal and peripatetic work at railway depots, as depicted in the movie
Lion, with whose ways they become familiar before undertaking their own
departure. But even in the context of the essential function provided by
railway space, most village children have access in school (or madrasa),
informal networks, village common space, and home itself to the story of
some runaway or other from their village. It cannot but be an active cultural
form that has developed in a complex calculus with — and in response to —
shared historical conditions.

Rural Calamity as Child Precarity

While some of it takes place in the city, this book is about a curious and
tender thread connecting the rural to the urban, and it is thus fundamentally more about village and countryside than the destinations and depots
to which children run. If the streets of Delhi, its temples and shrines and
squares, its tea stalls and festival grounds, as those of Bombay and Calcutta
and Madras and Lucknow, of Kanpur and Bhopal and Nagpur and Chennai, are full of village-born children who have left their homes, and left of
their own accord and without parental sanction, Nabil's cognates and compatriots, then the "countryside" is also full of the children who might become those "urchins" of cityspace, nascent street children sleeping amid
fields, beneath roofs of thatch and slate and straw, or no roof at all, but still
embedded, still nestled, in kin and rural circulation. They are not street
children yet, but any or many of them might be. These are the runaways
who form the subject of the stories of this book, the characters whose histories will fill its pages, and though this book's field of vision is often urban,

the aggregate movements it observes, the constellation of solo child migration, are profoundly and basically rural. Thus this book starts with a set of processes that amount fundamentally to questions of deep and drastic rural social change and of the notion of a "social fabric" in the countryside, which, if the notion holds up, might or might not be said to have collapsed, and with the question of whether such a thing, if it exists in that unitary, imagined sort of way, can "collapse" and "disintegrate" to create the condition that has so long been called "anomie." And though—or if—this is a historical problem of the agrarian, of the village and "countryside," and of the dialectics of these narrative and material formations with cities, it is rendered legible here in the lives of human beings, of individual children, of families, of groups of friends. The historical is lived, rendered material and phenomenological in "histories of the present" (following Foucault 1975: 31).

We might suppose that children are leaving home in response to the collapse of something, something they have perceived as having fallen apart. What it was that fell apart might have been the life of a place as the people living there themselves defined it, a remembered past, or it might be something more intimate, or it might be the former experienced as the latter. If the question is guided by what is locally and indigenously defined as *place* and *lifeway*, and furthermore as a *falling apart* that is legible as such a thing in emic cultural terms, and if the people living it are quite certain that such a thing happened, then it renders the question of anomie rather straightforward and simplifies our worries of romantic over-reads and externally imposed tales of woe that we author to satisfy ourselves. And indeed, this falling apart might to the person living it appear to be something of the largest scales—nations, markets, histories—or of the smallest. More likely the latter: what it was that fell apart in the end is more often *my life* or *my family* or the stability and persistence of *my child's being a child*. The tales are full of divorces, minor shames, an overwhelming fear on a single day, a fight, a debt, a job. But, indeed, where before there was something considered a fabric, there remains in its wake a perception of disaggregation, of something no longer remaining, of an order and a system vanished, on all scales. The lost order could be as small as what occupies a dyad, an intimate relationship. But something has fallen apart. The perception is spectral, phantasmic—one can sense it, but one cannot say what just came to pass, or why. It is something more than just memory and less than a ruin. It presents itself more as absences than presences, as presences that just were, one could swear it, but that have now melted into air and left behind only the faintest scent—a barely perceptible trace—of their former existence.

But the role of calamity in a more material, objective sense is worth exploring as well, as crisis is given form in experience, in life history, and imprints itself profoundly on children, as potential commodities, partial citizens, and nascent chattel—even slaves. I propose that stressors that lead to forced or voluntary child migration, and to periods of high child mobility among the poor, increase in times of flood, famine, conflict, and scarcity, for obvious reasons. The question to be asked is what shape such translations—between "collapse" and children *out* of place (though again, we ask who is it that decides what is a "child" and what constitutes "out of place")—may take, and how they take them. How might a year of storms turn into a rebellion that ends lives? How might a pathogen's sweep across the land pull apart a thousand years of local ties? How might a drought, or a pest, lead to increased landlessness, and how might that landlessness lead to family strife so vicious that it might cause a ten-year-old to pack his bags and take the first train out?

And so I ask here what kind of history can be uncovered in this part of the world of the nexus of child dislocation and crisis. How does calamity translate into predicaments, situations, and decisions that seem to be individual? The qualification must further be added here that events like famines do not simply "happen," and Amartya Sen's (1999: 16) assertion that "no famine has ever taken place in the history of the world in a functioning democracy" seems obliquely relevant, as crises, even natural ones, are situated in the matrix of wider—and often exploitive—social conditions.[4]

Of Blood and Garlic: Motives, Methods, and Anxious Ethnography

We anthropologists like to think about our subjective position in regard to the people we study among, about, and with (see Rabinow 1977; Geertz

1988; Clifford and Marcus 1986; or Crapanzano 1980 for classic statements about this). We still recognize a certain essential (if perhaps narcissistic) act in this reflexive self-examination, an effort to grapple with the bias and the relations of power we bring into the field and that enter into our relationships there, alongside the way we change, in a quantum sense, what we are observing by virtue of our presence. Early ethnography absented the ethnographer and, thereby, usually his (and indeed *his*) colonial identity and thus his association with a ruler. But the self is not the all, and sometimes, when we recount world histories or try to sketch economies, landscapes, languages, it is barely the *any,* and sometimes this "I-witnessing" (see Geertz 1988), essential though it is for an honest account and for our struggles with the question of who we are to speak for another (or to take up the legacy of a practice whose roots extend to empire's efforts at surveillance), is carried to the point of excess, and, occasionally, to fault. The self is indeed not the all of it. But still. A self is there. A self is in dialogue with these others. The story of the writer is, after all, part of the story. In a novel, that is kept quiet and left implicit; in a book like this, by contrast, it is made explicit. It is my debt as an author, I believe, to ensure that readers know who I am to speak for these children, who I am not, and how I presented myself to the children. On top of that, what it was it like for me, what my motives were in going, what I saw and felt—all these become integral parts of an honest account and, ideally, a complete story.

I grew up in a conflict-riven home of Tintins and *National Geographics*, which provided an easy escape from the humdrum drudgery of New York's haggard real and emotional grays. Later it was homes, rather than a home. At sixteen, flying unaccompanied, I was allowed in the cockpit of a British 747-400 while overflying India. It was night. It was before the Age of Terror. There were thunderstorms far below, and lightning flashed across a vast plain of cities and dark spaces that held tens of millions of lives I'd never even considered. That held my imagination captive. It still does. I had read Kahlil Gibran and Herman Hesse and maybe even Baba Ram Dass, and I had drawn a good swig of the exotic by then. I had to go. I had to find out what was in that vast and unknowable land. So by seventeen, I tried to make the *National Geographics* and the Tintins come alive by ar-

ranging a stay in a militantly pacifist ashram in Tamil Nadu at the dry eastern edge of the Western Ghats—where long curtains of monsoon mist from
the Arabian Sea streamed through gaps in the mountains but left the land
arid—run by my grandmother's friend, the late peace and literacy worker
and, eventually, honorary parliament member Muthukumaraswamy Aram,
a Gandhi disciple.

At Kovaipudur, where the ashram stood and still stands, I was disori-
ented and a bit traumatized by what I saw, and I felt confused, constricted,
and isolated. It was not the Hesse- and Schopenhauer- and Upanishad-
evoked land of misty ashrams and wise swamis I had imagined. Instead it
appeared to me a shocking land of excrement and dust, disfigurement and
death, and I could not bridge the gap between what I'd imagined and what
I saw. I'd imagined incense but was confronted by sewage. Tubercular and
polio-stricken children just a few years younger than me whom I met at the
hospital where Dr. Aram's daughter worked elbowed the gurus out of my
romantic Indian imaginary. And during this time I came to feel a strong
affinity for the dislocated children I saw everywhere, who were, after all,
not so far from my age, though so far from me in other ways, and many of
whom had stories of broken homes to tell. This frightened me even as it
obsessed me.

That summer, at seventeen, walking with a chaperone sent by Aram
on the long beach at the Kanyakumari that under the Raj was Cape Comorin, India's southern tip (everywhere I stayed in crumbling, mildewed
Gandhian rest houses and visited temples, even inner sanctums into which
foreign adults were forbidden entry; once I spoke contemporaneously,
poorly, at a Gandhian school near Madurai), I made my first acquaintance
with an Indian runaway, Krishna: he was my age, or one year my junior; I
wanted to befriend him, but my chaperone restricted our communication
and, troublingly in retrospect, kept the boy with him. I wanted a friend,
wanted his company, as I'd been around few others my age that summer.
That restriction, that lacuna, was formative, and I wanted badly to know
what had happened to the boy; it occurs to me now that still years later I
was pursuing a knowledge and understanding I'd been barred from gaining
that day. But from even earlier I can reconstruct memories in which I was

bewitched by a romantic notion of children alone. My first such memory is of a pair of Romani children busking on a French street whom my parents allowed me to approach to place a coin on their glimmering pile (*Daddy, what are they doing?*).[5] Quite a thing, I thought: I still haven't shaken the image from my mind.

Despite the shock of it, I resolved to return to the subcontinent, and quickly. I decided I'd find more of what I was looking for in its northern reaches, in contrast to the southern, for that, I reasoned, was the *true* land of the exotic, of Rishikesh and the Ganges, the Beatles and Ginsberg. With the help of a generous Palestinian American woman at Save the Children's headquarters, I ended up at the charity's Pakistan offices in Islamabad and Peshawar, even though I'd hoped for Nepal. There I helped to sort Afghan refugee crafts and write reports. At night, however, I walked out into the world and got wonderfully lost. Eventually, I found a group of Jāt children selling jasmine garlands on the street at night, by Nizamuddin Road in Islamabad, and thereafter spent some time every night in conversation with them.[6]

One night, out there by the crossroads, I encountered the jasmine seller Bilal, perhaps four years younger than me, weeping and groveling before a police officer on a motorcycle: no, he was saying to the officer. Please no.

I guessed at the situation and took a gamble, told the policeman, leave him, they're with me.

Who the hell are you? he said. *You* come for a ride, then, if—

I refused, told him I was an American, using what at that time was still a privilege of cosmopolitan prestige. He doubted me. I pulled my passport and he left, annoyed and unfulfilled. As he sped off, I said to Bilal (though I knew), what was that?

Bilal, drying his tears, inserted his index finger into a hole formed of his thumb and other index finger and moved it in and out in a gesture well understood across cultures. The other children laughed. Bilal laughed. What else could they do? Rape, in such lives, along with other subjugations and violations, is nothing if not routine. I felt myself quite the hero that night, a sense that was problematic and certainly misguided, but it was an inceptive moment that gave me a strong, if inflated, sense of a role to play, superheroic and grand, for children on the street, and an investment in their lives. *I rescued that child,* I boasted to myself, *from a policeman!* I thought myself quite fantastic, a kind of human-rights champion of the streets. But I learned, quite quickly, to be self-critical, even ashamed, of such notions in the four years of self-consciousness-promoting college that followed, quite appropriately, to unimagine myself a hero and to unimagine street kids as heroes. I was eighteen, the year 1993. But the memory of the evanescent, effervescent feeling sticks with me, just like the Cape Comorin Krishna, and just like the luminous cockpit.

By 1996 I'd spent five summers in South Asia, four of them in Pakistan, at which point I followed the street-kid trail elsewhere, ending up with two summers in Moscow (the first as a field researcher at UNHCR). There after each day of work I set out in the long light of summer evenings to trawl the subterranean passages of that recently Soviet city for new forms of misery (as are now well documented in Höjdestrand 2009 and the documentaries *Children of Leningradsky* and *Children Underground*) and cognates of my subjects of interest in a new place. Now I also had Theory at my side, nascent understandings of power in and as space: my new superheroics. What I did not yet have was the ability to problematize the *category* of "street children," itself. But I started to get it there: I found runaways in abundance, in those tunnels and station fringes, and orphans, subjects failed by the sudden disappearance of universal services, many of them Roma, and some migrants from the edge of Russo-Soviet empire. Some of the Roma were in unfortunate situations by virtue of nascent or timeworn forms of dispossession and racism; some were embedded in families and engaged in established cultural forms of begging. At that time a new group of non-Romani Central Asian "Gypsies," the Luli, Mughat, or Jugi, of indeterminate migration out of India (see Marushiakova and Popov 2016), had begun to show up in Moscow; a large portion of the actual Romani population was from Transcarpathia or Zakarpattija, Uzhgorod (Magyar: Ungvàr) in particular.

I watched trains from Grozny unload legless children, watched police interrogate people because they had brown skin or Muslim names. Indeed, because of my complexion, the police interrogated me. I formed a friendship in particular with a Transcarpathian Ukrainian Romani boy, whom I will call Kolya, who lived by the dumpsters at the edge of Kursky Vokzal's parking lot, cared for by a surrogate mother, herself homeless. Kolya was caring, in turn, for a puppy that slept on his mattress. The Romani "Gypsies" I met here, given their curious thousand-year trajectory out of the subcontinent, formed, for me, a productive link between South Asia and this new location. I refined my read of the complex cartographies of street life in Pakistan again in 2001, when there were still masses of Afghan street dwellers; in Dushanbe, Tajikstan (along with Samarkand and Tashkent, Uzbekistan) in 2003, where many displaced children, unhitched from formerly available Soviet services and orphaned by the war, walked the city and its new urban slums, alongside some more Luli-Jugi-Mughat; and in the streets of Delhi, which I tackled for the first time, properly, in 2005 and again in 2007.

When I accepted my current position, it happened that I was hired, as a special exception during a flush time, alongside another scholar of Islam. I was told that she would be covering most of the Islam courses, so would I please pick another area of interest in which to teach classes; I mentioned street children, the chair at the time said the students would love a course in it, and I jumped in to teaching the subject, rather blind. At the end of the course I set up a videoconference between my students, alongside local middle-school students, and former street children in Delhi. It was to be hosted by the State Department, whose server failed at the last minute, and so the CIA server hosted it. The local middle schoolers here, in the midst of war, started mentioning President George W. Bush; youngsters on both sides agreed they disliked him. The American videoconference minder in India squirmed and grew visibly uncomfortable; none of us knew it was the CIA that was hosting us, nor that street kids and middle schoolers can threaten national security. The videoconference received some local press attention, the dean mistakenly thought it was my primary research and nominated me for a grant that forced me to devise a project on street children, and my proper research on runaways was born. The grant proposal failed, and so did another on its heels, but I refined the project until I finally secured NSF funds to pursue it.

A figure of profound importance in this book, and in the work that constitutes it, is Khushboo Jain. In the research she did for this project, she is a full-fledged collaborator, and she is also—in part by virtue of her work for this project—the possessor of immense expertise on the fields of concern here, perhaps more than anyone else I have ever met. What she was able to find in fieldwork is central to and inseparable from everything I write about this.

Shortly after securing the funding from NSF for this project I sent out a call to a range of possible recruiting sites—universities, NGOs, and other sorts of institutions of advocacy and intervention—looking for what I then termed a "research assistant" who would work with me when I was in India and continue the work full-time when I was not. I heard back from a number of candidates; for a range of reasons, including her experience to that point and her enthusiasm for the work, I selected Khushboo. She had been working for one of the NGOs in question, and she knew many Delhi "street children," their stories, and their haunts exceptionally well. She possessed a remarkable linguistic ability, commanding English with exceptional fluency, itself certainly not rare in India, but over the course of her fieldwork she was also able to deploy rural dialects of Bihari or Bhojpuri to dive deeply into village life, alongside complete fluency in Nepali, no small thing when Nepali children figure prominently among trafficked, migrant, and runaway children in India.

What Khushboo accomplished immediately, and over the course of the four years of research, was stunning. I quickly saw that her initial job title could in no way describe what she was doing. I am certain no one else could have done what she did. It has, indeed, become part of her own work as a PhD candidate and advocate in her own right, a corpus she can draw upon for her own writings. Her accomplishments in the legal domain on behalf of railway-dwelling children has been truly remarkable.

I sometimes term Khushboo a "guerilla ethnographer." In her tenure on this project, she accomplished an astonishing amount of work, riding on local trains from village to village, traveling through zones of great strife, and braving the most emotionally fraught situations that can be imagined, villagers yelling across her at their children, children breaking down in states of life-shattering crisis. She saw child after child die over the course of this project. She advocated without pause for the children she knew and met. And to this day, as she continues major NGO-funded research on railway children in India and promotes their rights in courts and on streets, her resolve does not flag. A great deal of material in these chapters is thus the fruit of work done by her. She appears, as you will see, over and over again in these pages; her presence, indeed, suffuses them in interviews, interactions, and description. In many places, we worked together, and you will find her present in multiway conversations, on platforms or in shrines, that also include me.

Not long after hiring Khushboo, I left Vermont for India with my family again in winter 2011.

It is not as an expression of self-congratulatory pride that I say that street ethnography, or the ethnography of street life, is not easy, particularly when one is plagued by one's own fears, when anxieties and imaginings are filled with parenting, and when the emotional line between one's own children and such children becomes blurred, as at night, for example, when one cannot or does not fall asleep. In the course of our time in India I saw the newly dead, mangled, unattended body of a child recently run over by the train, as I will describe, and with Khushboo called the police that his corpse might be handled; elsewhere I saw the site, then embers and ash, of a child recently burned to death, a man at the Jama Masjid on death's door from what was undeniably AIDS, perhaps complicated by tuberculosis. As sores overtook his skeletal body and as he leaned over a vomit bucket, a homeless elder counseled: *the time has come to go home.* Sometimes missives can lend a distinctive sort of eyewitnessing, conveying one's feelings closer to the time they formed; to this idea the power of the epistolary bears witness. To a friend, in those days, one who knew India well, I wrote:

> I've been having an amazing but very very emotionally taxing time. Really quite difficult, in fact. Just walking through the smell of piss and death and rotting limbs every single day at Hanuman Mandir or Kalkaji or Nizamuddin Station has been a little tough. At least tough to do every day. Even three days

would have been a lot. So many sad children. And grim ones. And dead and dying ones. And ones with dead parents. And the glue rags affixed to their faces, and how they can't answer questions when they are high. And how I spent half the day today at a police station with a sobbing child (to whom the policeman returned the money, after all), and another with a head covered with blood (the work of an NGO functionary, no less). But that's life after empire, I guess, or life in a country that purges the street of such people for sporting events. So after a time I come to feel I can stomach it. And then, I suddenly think: why oh why *should* I be able to stomach this? Getting used to it is worse than needing therapy because of it. I'd rather be shocked than numb or accustomed. I am going to need some therapy. At night I dream bad dreams about my son, or I lie awake. I don't mean to dramatize. I don't mean to make heroics of it. It's just worse than I knew.

And, around the same time, to a different colleague:

As for ethnography . . . over time I have become distressingly accustomed to the substance of the work—a big hairy spider crawled out of a kid's hair the other day while I spoke to him in the police station; another kid showed me a knife-scored arm which he cut, he said, to "relieve tension"; a homeless man showed me in grunts and moans a photo of a very happy middle-class family (his own, before the accident that left him speechless). How can I get used to such things? It bothers me that I am, almost as much as it all bothers me itself.

The phenomenology of street ethnography casts its romantic imagining— wherein street children are exotic embodiments of authentic, ragged third-worldliness (now I'm *really* here, thinks the traveler on seeing the picturesque urchin)—into sharp relief. I must confess that I'd been pulled, myself, by that high heroism: the cultural reserve from which *Slumdog Mil-*

lionaire or even *Lion* draws its resonance, and into which indeed it adds stories and images, is deep and wide; it is part of what sells tickets to these films, what makes slum tourism so lucrative, and indeed what also draws donations, volunteers, and publicity to local and global charities.

But the experience was something else besides. I am an anxious person. I sometimes veer toward hypochondria, and the chessboard of hand sanitizers large and small I kept by the front door was not enough of a talisman to ward off the horrors I witnessed. Maybe I kept pathogens out, but I could do little to keep haunts, fears, and ghosts from my head. On the morning metro out of Chhatarpur I chewed on raw garlic, crunched coffee beans to keep the breath clean (until my family told me I reeked and that the runaways would be frightened of my stench). A tough, scarred teenager at the Hanuman temple caught the slightest flicker of hesitation from me as he offered his hand, having just coughed bloodily on it. You're afraid of me, he said, I see it, I know you are. By noon on any such day I was literally expectorating, alongside the fine particulates, other people's—multiple people's—fecal matter aerosolized in the roadside dust; I could taste it. Everywhere were gnarled bodies, sputum, blood, the smell of solvents, the presence of misery, the specter of death. The street was filled with a texture of sadness, loss, distance. I did not wish to bring it home. In the evening, when I returned to family, I quarantined myself, took off my toxic clothes, washed the day and its traces off before I'd let the little one hug me. Or so I thought—the washing did not work; the traces remain.

A Note on Poetics

The traces are manifest now in—even converted to—this writing itself, among other things. It is worth saying something here on just how I've chosen to write this book, and why I've made those choices. I mean this book not as a solution to a problem, nor as an exhaustive survey or scientific study, but rather as a meditation on that problem guided by questions, as I have said; I mean to plunge you into it so that you too might come out appreciating its complexity, its scope, and its sadness. The reader might note an unconventional style, or mix of styles, here: various departures from

conventions and norms of academic prose. This is intentional. I am inspired, motivated, and perhaps even emboldened by works inhabiting a
certain type of borderline between literature and social analysis, books that
tell stories compellingly but also rigorously. Some of them are novels, some
are ethnographies, and some are works of "nonfiction" oriented toward
questions of social justice (for example, Wilkerson's (2010) *The Warmth of
Other Suns,* or certain works by Barbara Ehrenreich). Among these, Agee's
(1941) *Let Us Now Praise Famous Men* is foundational, especially in its deployment of language; other works have of course done similar things for
India, Sainath's (1984) *Everybody Loves a Good Drought* prime among them.
Among ethnographies, I am motivated by the experiments in style and the
imaginative prose of Biehl's (2005) *Vita,* for example, Stevenson's (2014)
Life Beside Itself, Mol's (2002) *Body Multiple,* and Stewart's (2007) *Ordinary Affects.* Fadiman's (1997) *The Spirit Catches You and You Fall Down*
occupies an interesting space between popular nonfiction and ethnography. A bit closer to the thematic focus here are many works by Veena Das,
especially "Our Work to Cry, Your Work to Listen," in *Mirrors of Violence*
(1990), and *Life and Words* (2006), and Pandian's (2009) *Crooked Stalks.* I
strive for this work to embody some of these others' feeling, their adventur-
ous spirit, the authorial respect they endeavor to achieve as regards their
subjects.

Indeed, questions of how to write (i.e., represent—and represent
not only accurately but in a way that gives a story the gravity it mandates)
culture, truth, and reality constitute in anthropology a central and often
contentious matter. The "reflexive turn" of the 1960s (though properly
this "turn" began much earlier [e.g., Bateson 1936; Bohannan 1954] and
truly flowered much later), with which I inflect the who-am-I and why-I-
wrote-this-book discussion just above was concerned not only with how to
position the authorial ethnographer self in a matrix of historical local and
global power relations but also with how to mobilize prose in a truthful way,
where "truthful" is a matter of both expressing honestly and expressing
poignantly, which is to say in a way whose poetics somehow can be made
to match the visceral situation observed, emically and etically, somehow to
transmit its feel better than or in a different way from "scientific writing."

This debate on the proper role or nature of prose in academic writing thus involves an inherent and explicit critique of the normativity that prescribes a certain set of styles *as* academic writing and rejects others (as somehow less than valid). The friction between these various views extends well beyond anthropology and into the social and even the natural sciences. It includes debates on the degree to which writing based on "scientific" research should be constituted of storytelling and on how imaginative that storytelling should get. My goal in this book is to tell stories in a manner that also evokes and echoes the context in which the stories themselves happened; this can help express mood, memory, and the complexity of human experience. How do we convey lament or longing—or itch, smell, jitteriness, exhaustion? As affect and personal experience are such basic components of the social constellations I aim to describe, fashioning a book that tells the tale in a resonant manner is a critical imperative. I remain concerned throughout with the poetics of prose (following Jakobson 1987; and Bakhtin 1941). I am keen on thinking about how form plays a role in conveying content, as cadence, tone, and rhythm—a book's feel and its sound and the practice of writing are to me essential components of ethnography and cross-cultural research. They help express others' voices in a more honest and intimate way and give the reader a more visceral feel for fraught tales of suffering, sadness, hope, and loss.

On Gender

This has to be formulated in no small measure as a problem of gender. The runaways visible on the streets of India's cities are prevalently male, though many "street children" are girls. I am mindful of the fact not only that "street child" and "runaway" are constructed, publicly accessible categories but also that they are more often than not gendered constructs, by which I mean to say both that they are imagined, depicted, and narrated as properly—if not prescriptively—male spaces and also that they are spaces *for* male gendering and gendered subjection, in other words for the production of a certain modality of gender as it intersects with other elements of persona like age, kin(lessness), position in public space, aesthetic appearance, ability, skin color and phenotype, rural birth, and poverty. Moreover, and more broadly, running away may in part emerge fundamentally out of the interaction of political economy and certain features of the largely rural gendered production of boyhood that transpires at the level of home, village, and peer group. Running away might in part emerge from the position of young maleness in a larger matrix of cultural expectations, ranging from norms of obedience to caste or class mobility, at the apex of which is labor.

A focus on gender's role and gender roles, primary among forms of discipline and power operant in the runaway child's landscape, is critical here (see Massey 1994). It is indeed not only that a majority of the runaways we end up being able to *see* are boys but also that the public space they come to occupy as "street children" is more available to be occupied by boys than by girls. The sight of a male ragged child is more normalized, more routinized, than the sight of a ragged girl. Given the differential forms of socialization for boys and girls, from the village point of view it is less common that most village girls would view running away as feasible. Girls' and women's movement in and through rural and urban public spaces is, further, deeply limited by the same complex of rules that decrees that their proper place is at home with their families.

The dynamics observed by Huberman (2006: 66) are informative here; the "ghat girls" she observed in Varanasi, selling postcards and offerings at the river's edge, rarely remained in the public eye after puberty. A girl in public, she learned, threatens the family's honor with suspicions of impropriety and places herself in danger. There are girl runaways, but the potential length of their time on the street is much shorter. Girls are much more likely than boys to be immediately trafficked into sex work, domestic forced labor, or other forms of slavery upon or before arrival in the city. One girl runaway we came to know at New Delhi Railway Station, a teenager I will call Sita, was notably unusual: she seemed remarkably cheerful, healthy, and autonomous when we first met her. She had a "boyfriend" who was also a railway-station resident, the "Arshad" whose words are included later in this book. When Khushboo re-encountered her, almost exactly five years on, still there in the same place, she looked far more gaunt and said she had been "duped by too many men." The before-and-after photos are troubling. She now subsists by gathering empty water bottles at the railway station. But it is remarkable that she persisted there nonetheless.

Runaway and other "street children" in the city both are subject to gendered orders and exert their own gendered and age-bound forms of power—gendered even when they are within single-gender (which nearly always means all-boy) groups, because they hinge on the enactment and embodiment of gendered norms and roles—some of them enforced through rape as tribute and others as physical coercion, bullying, forced labor, monetary extortion, food-withholding, stigmatization, or simple expectations of normative male behavior like toughness. The axes of power governing formerly rural children's lives in urban space include social class, caste, religious community, physical ability, sexual identity, and age. In all of these categories the children of India's railways represent the outcome of some form of exclusion. The axis of gender is rather different, however, and requires some scrutiny. If we return to the fact of the prevalently male composition of populations of runaway children, it is hard to say whether presence in and claims on public space by boys represent an entitlement or a privation, a boon or a bane, or perhaps both at once. On the one hand, it could be said that these children are able to have the autonomy to be in this position of defiance and relative freedom, and to leave home (and thus, in many cases, thereby to escape subjugation), only because they are male; free living for boys is certainly more widely condoned (if not celebrated) than for girls, including in cinematic frames and other media. Indeed, it is maleness that makes a kind of trivial, carefree departure possible. When asked what sparked the desire to run away, one returnee from Sitamarhi, Bihar, said: "I like exploring, I hear about these cities from relatives and I just go to visit." Many male ex-runaways spoke wistfully of the unique (if temporary) freedom the city, distance from home, and separation from kin affords: On the other hand, it could be said that it is their maleness, and the social notion of their availability for labor and the expectation of money-making, that places them in the running-away predicament to begin with—subjecting them at length to its deadly entailments—and more or less compels them to leave home. Running away is in that case a realization of a sort of inevitable doom rooted in being male. Being a boy would in this view be central in what puts children more frequently in such terrible situations on the street. The now-adult former runaway Amir problematizes it like this:

> Yeah, being born as a boy is very difficult in countryside, so when you're born as a boy, your expectations, one's expectations get higher from you because they know that the boy is a source of income for them. They don't think about education, or how you're going to develop, or anything, so once you become, once like you turn nine, ten, eleven, then they start putting you in the fields, or whatever kind of work they do, they try to teach you, in certain ways a lot of kids, they don't like it.

Privilege or burden, then? Such predicaments over labor complicate even the construction of maleness, and its iteration in certain freedoms, in such a context. It certainly cannot be said that labor expectations and constraints for rural girls are any less severe; indeed, just the opposite. But to say that disciplines, constraints, and subjugations of gender engage both boys and girls, if radically differently, is of course true.

Running away is thus fundamentally gendered—in other words it is in dialectic with social structures of gender, and the construction of "street child" and "runaway" (and the indigenous renderings I explore later) generate a social space and an imaginary that is gendered in a very particular way. The street is also gendered, and violent in its gendering. At the same time, "the street" can potentially—maybe, sometimes, hypothetically—be or seem a space for certain freedoms of gender expression impossible in rural space, or freedoms *from* gender constraints in that space. "I really liked the freedom," said one returnee. "I could roam around anywhere and do anything." But these "freedoms" are minimal and evanescent; they are as quickly exploited or crushed or stigmatized as they are expressed or enjoyed. Such a thing might occur through trafficking into sex work, or rape, or coercive normalization, teasing, and more. Nonetheless, some children in India do run away *for* reasons pertaining to gender or sexual self; in the United States, clashes at home over sexual or gender identity and home life are a leading factor in runaway teens' departures from home (see Finkelstein

LGBT-identifying "street youth" shows off his *mehndi* near a Delhi temple

2005). And some runaways find the ability, once in the city, more openly to engage and explore a range of sexual and gendered subjectivities. I intend to highlight, here, those street-dwelling youth for whom such a thing is a core element of identity and life story, and not just the sexual companionship, intimacy, coercion, and hierarchy that are common among gendered groups of runaways. On occasion, runaways join or seek transsexual or transgender communities, including *kothis* and, notably, *hijras,* who hold a special form of ritual power in a range of South Asian societies.

Angles on a Journey

This book is about the interplay of historical forces and emotional constellations that converge, bundled or packaged in fraught moments, to make a child decide to run away from home, about what happens once she/he is in the city, about what home becomes once it is distant, and about what happens when the child returns or does not, or dies. I endeavor to broaden this question so that it is not simply about how history comes alive in a single child's life, or in a set of separate, disconnected individual cases, but in shared, connected ways in the lives of countless rural children by virtue of shared, connected, common historical conditions that plague the country-side. It is in the purview of this book to think not only about the children who leave but also about those who end up not leaving, would-be runaways who remain village children, who stay rather than flee from the conditions of suffering they are experiencing, or *toward* the conditions of aspiration and electricity—the pull of the city—they might otherwise be approach-ing. Some of this may disrupt assumptions about what constitutes a uni-versal or normal childhood, about the features we attribute to childhood in general. I allow much of this to remain in the garb of questions rather than answers. People are complex, and there may not be a single truth behind all this, but a range of complex, messy, loose truths. I follow as principle the spirit of an assertion by Elie Wiesel (1960: 4–5) that *every question pos-sesses a power that is lost in the answer.*[7] Good questions are perhaps more instructive and more illuminating than good answers, which fade, change, or fail. My work here is thus to provide different windows that shed light on the problem from different angles, glimpses that allow clues to what a com-posite, a mosaic, might look like. But we also must not assume that there is a single composite, a unitary reality that underlies all running away. Rather there are widely present conditions that people share from place to place, but of which they have radically different experiences, based on local con-ditions, culture and place, and idiosyncrasy. Global forces are infinitely di-verse in their various microcosmic incarnations.

I thus endeavor to build an approach that integrates ethnography and history, that joins or weaves them together in a synthesis in which the in-carnations of deep pasts in immediate, phenomenological presents can be called to presence, indeed where an archive sits comfortably alongside an interview, where a building's history keeps company with a quarrel's deep and polyvalent complexity. At the heart of my theoretical orientation is an array of ideas about how exclusion at the largest scales is rendered intimate, translated to the smallest scales—bodies, feelings, lives, cells. For that rea-son, I draw heavily upon the work of Paul Farmer (1996, for example), in his theorization of structural violence, and also from theses and proposals more broadly on global exclusion, all of them of course inflected to various degrees by Marx, from Sassen's (2014) examination of "expulsion" and the "systemic edge" to Sanyal's (2007) of "wastelands" or Li's (2010) "dispossessory dynamics." I look also, following Harvey (1985) and Lefebvre (1974), at the urban as a site of struggle that is inextricably yoked to rural catastrophe. At times I follow the thread of Foucauldian concerns of the child in the context of the carceral; the governing of the body, above all in "humanitarian" institutions; narratives of normalcy and abnormalcy in the construction of childhood; and disciplinary machineries and dynamics in the management of urban space. All of these are manifest in historically specific forms, discourses, and technologies that are not at all abstract but rather translate directly into children's affective, visceral firsthand experiences.

I approach all this from a carefully considered range of conceptual angles and spaces. Initially I had planned a narrative that moved literally and linearly through a chain of actual nested physical spaces and scales, or space and scale types: village, home, railway station, city street, institution. I've kept something of that, indeed, but I also have found that such "spaces" are often not discrete, and not always spaces alone, that they are intertwined, entangled in a knot of other places, and that the processes and stories of running away are too messy to box them up strictly by spatial taxa. This is to say that the places are materially and socially bound together, in terms of real interactions and flows, and also conceptually woven, in terms of similar conditions and structures. Moreover, different places are woven in complex ways with different *times,* which themselves deserve their own attention. In other words, particular places in the cartography of running away cannot reliably be privileged to the exclusion of others, or in isolation. Various places and times are in, at the very least, dialogic or dialectic relationships, if not vast and expansive complex systemic relationships — polynodal matrices and web-like lattices, perhaps, if we had to map them, networks and mazes incarnate in multiple dimensions of time, space, and experience. Running away is not a single thing, and neither are its spaces or experiences. This means that even to a single child the act of running away need not be aggregated into a unitary process, simply because we impose upon it that interpretation, and it also means that running away is not necessarily the same thing case-to-case.

It should be said that the observation about the imbrication of (what appear to be) discrete places with each other, and with other forces, forms, and times, is true both of the material and social lives of the places themselves and also of the way people talk about them, their narrative construction; the actual lives of places are tied up with idealized imaginaries of what they are; those imaginaries are themselves, in what Bakhtin (1986) called *chronotopes,* tied up with concerns about time, through memory, nostalgia, or aspiration, for example. So what "a" or "the" village is, what "home" was, what "the city" was before arriving in it and what it is now, all these are situated in nonstatic, charged, and complex narrative constructions of places *as* times, experiences, and symbols of something. Not only are such narratives complex and nonstatic, but, as with any sign, as Vološinov ([1929] 1973) points out, they also are deeply contested by the children, and also in larger political fields.

There are fundamental statements in this book to interrogate the very categories on which the analysis is built, in particular of the "runaway," of the "street child," and thus, fundamentally, of course also of "the child" and "childhood" themselves. I am indeed at times in conversation with a small but robust corpus of studies of "street children," among whose primary goals is to challenge popular, institutional, and even scholarly constructions of that category of subject (for example, in Glauser 1997). Some of these focus on destabilizing the notion that "western" views of childhood, including scholarly and psychoanalytic ones, have anything in them that can be universally applied (or universally understood to be features of childhood). The extension of this argument (including in Heywood 2002; and Cunningham 1991, 1995, and 2006), which I find quite effective, is that perhaps childhood as a phase of life is itself a cultural construct with roots in historically specific class-based ideologies. Such ideologies have institutional entailments and thus iterations, for example through charity. The childhood-as-construct proposal is not without its detractors, most notably David Lancy (2008), who argue with similar vigor for a certain irreducible notion of childhood as a discretely bound phase recognized by all or most cultures.

The lives of street children are brought to bear on such questions through the mobilization of cultural relativism, in such works, by suggesting that in other cultural settings "street childhood" might not be seen as anything shocking or out of the ordinary whatsoever, which is to say that children alone on the street could be simply manifesting a different view of youth and life course than is expressed in the perspective of those who cast "urchinhood" as an aberrancy to be fixed. This book differs in important ways from many or most of those, for example in its fundamental engagement with the historical landscapes of varying scales in which solo children can be said to be embedded.

Nonetheless, though this work does not necessarily emerge directly from that corpus, and though it is not primarily a refutation of the idea of "street children," it must nonetheless be positioned in relation to their proposals and accomplishments, and in the context of the work they've done to question the very idea of "street children" as an idealized and unitary form. In particular, I embrace and deploy the critical observation that street children are, as a visible public category, in part constituted or constructed by virtue of their violation of normative orders of who should be where when, of being in or out of place: "poor children in the wrong place," as Scheper-Hughes and Hoffman put it, "governed by the preoccupations of one class . . . with the proper place of another" (1998: 358). As Glauser (1997) notes, the ways that children interact with city streets "contradicts not only dominant ideas about situations suitable for children to grow up in, but also ideas about the purpose of the street or any public space seen from the point of view of adult and class needs and uses" (152). Hecht (1998) observes, similarly, that "street children challenge the hierarchical worlds of home and school and threaten the commercialized 'public' space such as malls and shopping centers. They subvert their country's unmentioned but very real social apartheid that keeps the poor cooped up and out of view" (211). In my chapters on railway space, on urban cleansing, and on street children's engagement with death, such observations form a crucial starting point for the stories I tell.

In chapter 2, I describe the quotidian and local conditions and features of the homes and villages that the children leave, and, through a few individuals' stories that I trace in depth, I introduce readers to the idiosyncratic diversity within and structural connections between children's departures and trajectories. I conclude the chapter with an exploration of what happens to "home" in the longer range, over time and when children return or do not. In chapter 3, I add historical dimension to the previous chapter's more contemporary, ethnographic angle on the village and departure, looking at deeper and older forms and forces associated with child precarity and departure, from agrarian debt to legacies of slavery. I endeavor to underscore how the past comes to life, almost invisibly, in the present. I continue in chapter 4 with a reflection on the children's relationships and encounters with death; following on the previous chapter's meditations, I examine, again, how death—and susceptibility to death—are interwoven with global and local histories.

Sometimes that death happens at the hands of the railway, or in railway space, and the public's image of where poor children die is itself also imbricated with trains and tracks. Chapter 5 is an exploration of the children's engagement with and presence in railway space, a theme depicted, but of course not thoroughly unpacked, in *Lion, Slumdog Millionaire,* and beyond. Children use the railway to leave home behind and get to the city and often stay in railway space for their whole sojourn in the city, or indeed for their whole lives; it is the thread yoking village and city. The railway constitutes perhaps a more powerful metaphor, rendered brick-and-mortar, than any other for child runaways' intimacy with history's forces— empire, capitalism, and rural transformation among them. It is also a space for a very vigorous control imposed upon children's bodies and movements through the vehicle of the state, of informal economies in global capital, and of other mechanisms of power, just as it is a space that the children in question *occupy* in a type of evasive practice that is irksome to society and government.

Chapter 6 examines the thin boundary between the charitable and the carceral embodied in the institutions that both aid and confine runaway children. I unpack this thin boundary both synchronically, by instantiating contemporary nongovernmental organizations' constructions of "reform" and "rehabilitation" and considering their complicity with campaigns of urban cleansing and with structures of policing and confinement, and historically, by excavating with archival research continuities between extant and antecedent charities for "vagabond children" and the colonial reformatory itself, particularly as it was applied to the children of societies that were constructed as criminal-by-birth. In my concluding chapter, I ask three foundational questions to explore what is at stake, intellectually and fundamentally, in the patterns that can be discerned in children's departures across India. This chapter situates the book in a corpus of theory that informs it and with which it is dialogue throughout. The first deals reflexively with the book's project itself, asking: what are the ethnographic ethics of researching, depicting, and describing runaways' suffering for academic or artistic ends, and what are the entailments of speaking for someone else, particularly a vulnerable, nearly powerless subject? The second question asks whether running away might represent a form of resistance, and if so how, and to what power. The third question situates running away in the context world history, world history theory, and architectures of global capital and labor. These questions help frame, define, synthesize, and sediment the book's wider relevance at its end, at the point where the reader can properly take stock of the story from a rich range of angles.

T W O

Home and the World

Village and Child Up Close

Leaving Home: Variations and Themes

I wish to start with some guiding questions that lead right to the runaways' own voices, stories, and experiences, to begin with an interrogation and exploration of the village and the child, rather zoomed-in, taken from a closer range. The scale of this book should be both intimate and grand. The runaway's journey illuminates something of what it is that *connects* the intimate and grand, in fact. My intent is to move outward from the *origo*, the point of departure, to take stock of a certain type of journey from start to finish, to take its measure, to make sense of it, to discern its form and shape, and to see it as clearly as possible for what it is, if it is indeed a single thing. *Is* the journey a single thing across place and case? It seems to be. Why else would so many children do the same sort of thing and narrate it in just the same way—and without much access to information about the others— without some widely shared structures that make it possible or ideal? Or is there, by contrast, information passing in ways that none involved can describe or even perceive and that thus goes unremarked? And if it is a single thing, how is it, and why? It is worth asking whether it is also a single thing *for* the person involved, or an aggregate of different experiences onto which we—or others—bestow a unity once seeing them on the street and deciding that a single journey brought them here. Maybe it was many disparate pieces of a life.

It is paramount here, in the midst of theories and longer histories, to center things on the children who form the heart of this inquiry, on who they are and what happens to them over time. From several peculiar vantages I will look at this or that aspect of a runaway's trajectory, but first I attempt to let the idea of running away and of the runaway child come into resolution from a more proximate vantage that will reveal something about what it looks like zoomed-out, to permit it to take shape, piece by piece, as an entire map that has a form and reveals pattern. Such pattern could come into focus first in a region, and then across regions in a larger domain (like nation or subcontinent), and then potentially across the world, among areas sharing certain historical characteristics. It is my supposition that pattern does emerge in comparisons on all of these scales, despite profound cultural and situational differences, and that it emerges in part by virtue of similar local reactions to global forces shared across the planet by virtue of capital and empire; "street children" and "runaways" take on similar characteristics in different places by virtue not of some human behavioral universal but rather of prevalent conditions, determined by markets and cognate pasts, across the planet.

I want to ask, what is the life of these places, homes, families — the places from which the children run away, the same ones sometimes haunted by the ghosts of troubled pasts — in the present moment, and to assert the primacy of the question of just who these children are. I wish furthermore to underscore and emphasize the constellation of circumstance, structure, and constraint in which children *make a decision,* despite all the forces and norms that decree they cannot or should not, to take their bodies and futures into their hands and leave their families and natal homes. Is running away resistance, or is it desperation, a response to coercion in the face of no choice at all? Is it migration? Is it in all cases an available pathway *a priori* that the children simply follow without as much autonomy and agency as I assume?

I posit that runaways are fleeing not so much the misery of poverty directly, but something a little more oblique and indirect: a stress placed by historical poverty on families and *translated* into abuse and emotional strain. But are these villages really such zones of misery? Are they so bad that fleeing to the city could only make things better? I don't think so, necessarily. I hold to the view that children who run away engage and mobilize an available model for possible action—indeed, for the taking of one's fate into one's own hands and for changing the terms of the control and direction of one's life. Here a child wrests that control from its earlier and normative locus, the home, the family, the natal village. This is counternormative first because it is not conventional for the *child* to assume such powers, in the eyes of society, and second because it puts the child in a situation and gives her powers that are not seen as fitting of childhood as publicly prescribed. Even locally, *within* the context of a "short childhood" or something we see as a childhood that is not in its context considered a childhood (Ariès 1962; Cunningham 1991; Hecht 1998), it is not normative: the twelve-year-old "non-child," if that's what he is in a certain place, is to be working and supporting the family, but at home. In leaving as *runaways,* children usually also remove their earning-contribution potential from the household, unless they were specifically sent away to work. If we consider, as we will later, the language of charitable efforts for children from the Raj to the present, we see that norms as are being defied here—and defied by the child herself—are laid out forcefully in practice, in law, and in what campaigns aimed at the engineering of the child's rehabilitation suggest about what childhood should contain and constitute.

On the surface, the texture of any imbalance to which running away is attributed—of lives, of places, of times of history—nearly always seems intimate and microcosmic. Take, for example, this narrative between me (J) and Khushboo (K), on the one hand, and the older boys Arshad (A) and Vikas (V) from Hazrat Nizamuddin Station, on the other, on the occasion of the boy Petu's death beneath a train, a story I will recount later:

J : Did you run away? How old were you?

A : Five years old, in second grade when I was five.

J : What about you?

V : I came in anger.

J : From Himachal too?

V : No not from Himachal.

J : How did you know that you could take a train?

A : My house was as close to the station as this station is to Paharganj . . . my house. I just thought of taking a train and going a little ways. I didn't know that the train was going to drop me at Delhi. It took me a day and a half to reach here. I got down from the train, I got out at the station, there was a water seller and he offered me some water and some *chhola kulcha*, and I started collecting bottles and studying a little bit with a very good *didi.*[1] I did some good studies and then that didi stopped coming. I kept collecting bottles and collecting bottles and then I started to feel ashamed about that act.

K : Weren't you scared when you left home, being so young?

A : What fear? I came because I was angry. I didn't leave with any idea of whether I would return or not. I just came out with the idea that my parents would get worried, look for me, and then I would eventually return. I thought that they would become frightened for me and stop beating me and love me. I left home with this thought.

J : Did you ever go back?

A : No I never went back.

J : Have you spoken to anyone else?

A : I have a brother. He lives in Delhi.

J : Do your parents know you're alive?

A : My father is dead.

J : When?

A : My brother told me that my father was death.[2]

J : Were you sad about that?

A : Of course I was sad.

Children, in general, as we see in that conversation, and as discussed in the last chapter, run away from what is on the surface an untenable family situation—a wicked stepparent, a family in crisis after one parent dies, undue punishment for labor deemed unsatisfying, physical harm, sexual abuse, shame at some act. Sometimes it is not narrated as a crisis at all: the child wishes to make it big, to star in a film, to be a gangster. Nonetheless, in these cases, the pull of such things overrides the pull of family, and leaving seems relatively easy—and normal. It is the larger patterning of these intimate crises that suggests a macrocosmic set of agrarian conditions motivating child departure and faltering spheres—faltering in the eyes of the subjects themselves—of kin and care, often indirectly by virtue of labor's strains.

In the next chapter, I will ask about the historical antecedents of child departure—antecedents, not causes: earlier forms that live on in memory, talk, or structure, as phantoms of a certain sort. Could legacies of slavery and child selling, of indigo labor and debt servitude, translate through structural means into intimate suffering in the here and now? How, exactly? By what mechanism? These are questions about the past's specters living on in ways that people cannot see through long and entrenched forms of exclusion, expulsion, and misery. But here I grapple with a question of the "history of the present," of what it is like in this moment in these places, who the children are who leave, and who they become in the city and over longer stretches of time rendered legible on the scale of lives and even of genealogies from generation to generation—but crucially not on the scale of histories that are not readily available to them now as causes of things in their daily lives. As an ethnographer, my impulse and tendency is to prioritize the view from inside, with the way things look locally, on the ground, to someone living them, not just to me or to you. There is a place for what I've been calling the zoomed-out macrocosmic view, a value to that analytic distance, but here I privilege the child who leaves, the people and place he leaves behind, the people and the place he leaves them for, and the weight of going back, or not.

The basic problem stems from the fact that so many children come to the city so concentratedly from such a limited set of places, and those places are not the poorest places and the families they come from are not the poorest families, but they are nonetheless usually very poor, those families and places—those interviewed ranged in ascriptive identity from Chamar Dalit

"Untouchables" to high-caste Brahmins (one family, caste notwithstanding, owning but a cow, a goat, and "a small house in a crowded place"), from poor Munia Kisan tenant farmers to landless fishmongers' families (with nine children), from shopkeepers to "Kabadi" waste collectors of the "Sunar" caste of goldsmiths. In a survey (2004–5) of 1,041 "home-placed children" (i.e., returned to their homes) by the NGO Sathi, only 12 percent were classified as "very poor." By contrast, 43 percent were in the "poor" category (but not poorest-of-the-poor), 20 percent were in the middle category, and 1 percent—11 distinct cases—were classified as "rich." A separate survey of 1384 children (2005–6) had similar numbers: 11 percent very poor, 48 percent poor, 38 percent middle, and 3 percent rich.

What is it about those places from which runaways depart, if it is not straightforwardly poverty, that makes them such dense sites of departure? This book ranges in its focus from the railway to traditions of mendicancy, across a wide range of forms that may interact with running away; the goal is not necessarily to settle on an answer but to identify operant and relevant forces, patterns, and structures. At stake here, among other things, is the role of transmission of talk, of knowledge sharing: *this is something you can do, you know, when things get really bad.* Kids talk to each other; sometimes, this manifests in sheer numbers: in Rasalpur Block of Dumra, in Bihar's Sitamarhi, for example, over fifty children were reported to have run away; in Haspura (Aurangabad District, Bihar), one now-grown former runaway Sachdev recounted that fifteen to twenty children ran away all in the same period as he did. When a railway station is nearby, it usually has its own resident kids who might provide even more education on what you can do and how you can do it. But we will also see here that sometimes there is no real antecedent, and that the runaway is an innovator who actively creates the antecedent.

But even if children talk to each other, villages separated by vast expanses do not. It doesn't explain why children seem to talk to each other in the same way over and over again in such similar ways in such disparate places, unless we believe in a kind of inevitable and fated structuralism where such a thing was always going to happen in all places that share certain circumstances, given those circumstances, and then manifest itself in talk and quotidian thought. Children leave from a limited set of villages, but those villages are not necessarily adjacent to each other and often they share little in common. There is, however, another kind of more easily historicized structuralism, another way models for action avail themselves, akin to talk: kids see movies. In some of those movies children leave home and make it big in the city. Consider this response from Amir, a former runaway from Bihar who is now an adult:

JONAH: Yeah, like how did the idea strike you to do this?

AMIR: I think it's like, I would say like movies maybe influenced it, less like I got into this situation where I felt very neglected by family, by society, by, I don't know, anyone I knew at the time, and felt that nobody loved me and so, like, "what's the point?" I'm going to go somewhere, you know, get out, like the feeling, of being separated from them and also then, you know, when you have such a feeling you want to go somewhere you want to make sure that's how you're going to survive.

Amir describes the decision as emerging in part from models available in film and in part from enacting an idea of what he wanted to do or be. Children's decisions are not only the outcome of an objective calculus of impulse, deliberation, action, and consequence, as empirically observed, but they are also motivated by *imaginings* of what they are doing, reflexive and recursive metaenactments of their own stories and of what they want them to be. Children see *Slumdog* and then imagine themselves as Jamal. They enact what Avery Gordon (1997) poignantly calls "complex personhood," which "means that the stories people tell about themselves, about their troubles, about their social worlds, and about their society's problems are entangled and weave between what is immediately available as a story and what their imaginations are reaching toward" (4). The flight to the metropole is heroic and full of longing and possibility. You might end up being better than this one-horse town decrees that you'll be. You might make it big and prove everyone wrong. You might be a movie star, a magnate—and, indeed, such stories are available and known.

Is there a typical child? A typical situation? A typical village? Possibly: it is my supposition that the power of the sharedness of structures that are incidentally shared among places by virtue of common historical conditions cannot be overemphasized—the railway and its associated conditions, the hopelessness of the smallholder, the pull of the city for labor as an ever-present trope, the shared *notion* and *discussion* of misery on top of the actual misery, and the hands (and other parts, and tools of pain or confinement) lurking in every way—to hit, to have and abuse, to take and subject, to enslave and take away. The typical village seems, if nothing else, aware of its plight, aware of its bad-placeyness, unsatisfied with its lot. It is susceptible to natural disasters, flood and drought in their eternal twain cycle, but it's not really that: it is susceptible rather to helplessness in the face of those disasters, it is precarious, it has no recourse to anything beyond its own weakened structures. The resources it has are close to its last, but not its last. The very poorest villages in India, near the intersection of Orissa, Chhattisgarh, and Jharkhand—especially the infamous KBK districts, Kalahandi-Balangir-Koraput—are plagued by profound violence between the state and Naxalite rebels, and ravaged by the exploitations of mineral extraction, but produce few runaways, perhaps because of tight kin connections in "adivasi" societies, such as Munda, Gond, or Dongria Khond, where cohesion is seen as imperative. Likewise elsewhere squarely in Chhattisgarh—Dantewada or Bastar, for example. In all or most such places on or near the Nagpur plateau, in the greater Gondwana region, displacement and dislocation are collective; they happen in or to families or settlements, and they also occur at the hands of one collective or another. Of course, state purges, the witch hunts of a proxy "people's militia" created by the state (the Salwa Judum), and other actions in response to Naxalite insurgency have occasioned a mass dislocation: southern Chhattisgarh and northern Telangana are sites of huge camps of internally displaced people. Such forced movements fracture the kin-based or ethnic collectivity that constitutes the initial shape of the fabric, the thing that has to exist to begin with in order to fall apart.

The landscape is different indeed in the industrialized, canalized, railway-etched Ganges plain, with its old patterns of land tenure, debt, and communal relations. The confluence of forms that both subjugate and empower, like the railway, and the weak cohesion and vigorous manipulation of cultural assemblages that tend to bind people tightly to place, like tribal groupings, make such villages, on occasion, rather difficult places to grow up, and the expectations for labor from children, both at home and from an employer (or the risk of getting trafficked or even being trafficked at the hands of one's family for money) are substantial and sometimes stressful.[3] The tribal, cultural, and ethnolinguistic connections that have more vitality elsewhere and that are all embedded in kin are in many cases less robust, and identities may be more thoroughly couched in caste categories (see Dirks 2001) that are fields for exploitation rather than cohesion and sticking together. These villages further reveal profoundly unequal structures of accumulation and distribution and often suffer from graft and extortion and new forms of debt servitude at the hands of politicians from hamlet to district. Such conditions of rural desperation were described (and indeed rendered emblematic of the Indian countryside) by Palagummi Sainath (1996) in his classic account of village misery at the hands of debt and "development," *Everybody Loves a Good Drought*. Aravind Adiga's (2008) Booker-winning novel *The White Tiger* is also in no small part about this. Culturally speaking, local ideologies in such villages do not invest children with particular self-realizing, voice-elevating power, even while childhood is not something considered in such places to last particularly long or to be very distinct from other phases of life. The decision to run away seems to emerge from the milieu of children themselves, not from any imbuing of autonomy by adults.

The external perception of an "urchin" involves abject poverty, desperation, and helplessness, but the typical child of this book, if she/ he exists, is also one who in the village setting is not entirely without resources, emotional and material alike. Amir, in recounting his story, affirms this sentiment: "I was a smart, smart kid, I've always said, even though I was the smallest one, but my defense mechanism was very strong—to cope with anything. When I get the sense of getting hurt, or that somebody's betraying me, I always listen to my gut feeling and always take a step before I take any steps. I always did things, I always took actions before they would take actions." The child who makes it by night to the regional depot and by day, and over days, to Delhi or Mumbai, has made a plan, done some research, consulted people, weighed things, and made a very tough decision crucially involving independence. This tends to be a child, as Amir was, of strong and decisive will who has assessed the situation in which she finds herself carefully and come to a point that requires rather strong resolve, and then, further, has taken an action that rends her from her natal hearth and settlement.

Typical kid, typical village: maybe or maybe not. But there is more unequivocally a typical *situation*. That situation tends simply to involve a stressor, and that stressor tends to be abuse at the hands of an elder or sometimes a peer. In the NGO Sathi's exhaustive records, "scolding" or "beating" appears in most surveys as the most widely prevalent cause of departure, coming in at the #1 position in the Sathi survey of 1,041 children as the reason for departure, and overlapping extensively with secondary variables of "stepparents" and paternal alcoholism. The antecedent stressor, of course, has its own sparks and catalysts; abuse, in many cases, seems tied to economic strain. Consider how Roshan framed it in conversation with Khushboo (as K):

ROSHAN: Father used to beat me, that's why I had to leave. The only people I told were my younger siblings.

K: Why did your father beat you?

ROSHAN: He used to not work. He would wake up in the morning and beat me. He would take all the money from me. And whenever I needed money he would not give any to me. Because of this I was not able to buy a notebook for school. My teacher used to beat me up for not having a notebook, I had to get my friends to study.

K: Your father did this to you or all your siblings?

ROSHAN: Only with me.

K: Did you ever talk to your father about his behavior?

ROSHAN: He's crazy. He does whatever comes his mind without thinking of anyone. He would stop going to work for days and just sit at home. He keeps borrowing money from other people and does whatever he wants with that money. My mother asks him to stop but he beats her too.

The elder may be a parent, an uncle or aunt or stepparent, a teacher, or some other authority figure. The abuse may be sexual or violent or verbal, or it may involve untenable expectations of labor, as we will see in particular in Firdaus's story below. Sometimes, as in the narrative of Ajay's mother, below, parents will defend their punitive practices as well-justified, morally-defensible disciplinary practice.

As parents we need to make sure the kids are on the right path. So it becomes important to scold and beat the kid, otherwise they don't listen. Ajay doesn't like his father because he beats and scolds him. But it should not work like that. Ajay should understand that his father does this for his betterment. He should not run away.

In such cases the child's departure itself serves as resistance or objection to the proposition that such sanction is reasonable. Indeed, given the absence of any other assets, the mobility of a child's body may be his or her *only* way to make such an assertion, statement, or objection, to engage in dissent. They cannot fight back, or apply for a different family. The strain of abuse from which children flee can also emerge, indeed as it does above, in Roshan's case, from a nexus of school *and* home, working in what seems to the child to be a concerted effort. "My parents," reported Anjeel from Roorkee, a city of the floodplains of the Himalayan state of Uttarakhand, "wanted me to do well in school, but I did not understand what my teacher taught me. So I was being crushed both at home and school . . . at home, my parents beat me, at school, teachers beat me. So I used to stay outside." Ashok from Motihari was beaten, before leaving, by a teacher for forgetting his tens times table, and found no sympathy from his father.

At times the abuse from which children flee can also involve captivity within a family situation, or captivity by virtue of being sold for labor. Sometimes, children flee a dispute, a feud, or an accusation. They may flee a nonsanctioned form of love or affection, sometimes with the lover in question, or the shame of a pregnancy. They may flee because of a non-normative gender or sexual identity, or because of bullying. They are, after all, largely eleven to thirteen when they leave, on the cusp or the onset of adolescence, and they are figuring out how to configure new realities of desire, gender, and autonomous self in the world. However, it is crucial to remember that running away often involves a pull rather than a push—the movies and tales mentioned just above, the hope of more money, a promised job, drugs, sexual opportunity, the excitement and pace of the city, the urge to find similarly departed friends, and the lure of services and shelter safer than homes at NGOs or shrines. Money, of course, looms large among these competitors: "We had been planning," said the now-grown runaway Mohan of his flight in 1994, "to go to Delhi. We *really* wanted to go to Delhi. We had heard that there's a lot of money and opportunity there." His assessment, from the vantage of his own village, was that "as long as a kid is working and sending money," parents are at ease with their departures. At times, parents may indeed even send their children to the metropolis, in which case it is not running away, though the public might not differentiate. One child (Ashok) claimed, confoundingly, that his running away consisted of his father sending him to Delhi to work, but "without his approval." And only *then* did he run away, right to where his father had wanted him to go. Omar Mullick's documentary *These Birds Walk* sheds light on the lives of a group of siblings in Karachi whose parents feel they are better cared for in the homes of the Edhi Foundation, a Muslim humanitarian network, and thus they send all of their children—who sometimes favor, in their own accounts, fabular running-away rather than true getting-sent stories—serially to live there.

Amir's Journey

On the question of a typical situation: the first case I want to feature at length in this chapter on children, village, and departure is the story of the Amir I've just mentioned, a case to which I returned five years after I first

had a glimpse of it in Delhi. Amir is, again, a grown runaway. The interviews with him were in a mix of Hindi and English, both of which Amir
speaks fluently. I include some of its first telling elsewhere, from an interview in a café in Connaught Place, in my chapter on runaways and the railway. When I was able to speak to Amir again in 2016, twice and at great
length in both Hindi and English, he was living in the United States. I heard
a story that went deeper in every way, that covered more emotional and territorial ground, a much fuller, more human, more complex, and more troubling picture. In the consideration of "what is typical" here Amir's story
reveals something both about what might be identified as "typical" in any
given runaway's tale and about what might always be open to variation,
what is fluid and what is fixed and how to listen for such things, and what
still hurts a decade later, or forever. Amir's story allows for reflection on
how the experience of running away ferments, crystallizes, and sediments
itself in memory and narrative, what it becomes in a life. Amir's tale thus
gives us intimations on the runaway across a longer expanse of time, another critical question in this chapter. It also reveals something about the
tremendously complex calculus between person, family, and village that
precipitates departure. How do place, communal identity, class, and history come together to impel one child to run away? I provide a long stretch
of Amir's story here raw and largely without voiceover commentary, which
all this already is: it is remarkable and speaks for itself and speaks well to
the interplay of individuality and structure in running away:

    JONAH: So, like, what's the thing that made you leave and
      how did you know that leaving was something you could do?

Here I (Jonah) was fishing for something that might indicate peer flow, a
culture of running away, but what I got was something much more idiosyncratic, if halting, a first nod to the presence of a disciplinary stigma in decisions to depart.

    AMIR: Oh, ok. So the thing that made me, and I told you, I
      became one of the bad, I mean, somehow, you know, I wasn't a

good kid, I was not considered a good kid in my village because I always fought with my friends, I always fought with family, I fought with brothers, you know, and I got into stealing, I . . . my friends were the bad kids who'd never been to school, did nothing.

JONAH: Yeah, so you became a bad kid or somebody else labeled you as a bad kid?

AMIR: So I mean, people started labeling me, so I kind of started drawing towards that, like look if you want to make me bad, I'm going to be more bad. So my friend circle became very, I had a very bad friend circle, they started going somewhere, like breaking into houses or like stealing fruit or vegetables, just kind of bullying sometimes, but you know I won't do it but those were the sort of friends I had and I kind of got a pretty bad name and then later got really mad and they wanted to give me a really bad punishment, and I didn't know what really bad punishment was [i.e., "meant"], you know, it could be really serious, like maybe kill or death, I don't know what they could have done, maybe just break something. So that's why one day I got just caught doing something and they tied my hands and legs and they put me on the roof and the roof was pretty, you know, hot, it was almost like, I don't know, eighty degrees or something and I was just burning, burning and then I realized I was in trouble, I need to get out of here and somehow I was able to remove myself from the heat and I was able to run away from there.

I wondered now if such a personal stigma could be tied to a larger one, especially since Amir intimated elsewhere that the idea of leaving had already hatched in his head:

JONAH: Amir, do you think some of that could have been from, you know, you've talked about maybe two-three years before you're already realizing that, you were stigmatized because you're a Muslim, like do you think somehow that stigma led you to that position?

AMIR: Yeah, because also I was a Muslim kid so that was definitely a bad thing for villagers because, you know, I mean my dad also tried to maintain his image and I was one of them you know not helping him out and just putting like bad image in the village so they definitely saw me as a thorn, and especially being Muslim, so that could be a reason.

JONAH: Your dad tried to maintain his what?

AMIR: Maintain his image, maintain his image in the sense, you know, he was a Muslim, he was a well-known Muslim, he always, people knew him for good things not for bad things, and I was the bad son for him who was spoiling his image.

JONAH: So who tied you up? Who did that?

AMIR: Villagers . . . villager people tied me up. So what happened, one day, I think me and my friend we stole something, so they tied us up, I was anxious, like I, I got the feeling they were going to cut something off, or just like, I have a pretty bad feeling that I'm, they're either going to kill me or they are going to do something really bad, and like my family, all these people, were watching from the roof and just when my mom was just really crying or upset or whatever, and I somehow got the feeling that I have to get out, I have to run away as fast as I can, before, you know, it gets too late.

JONAH: Do you think this would have happened if you were a high-caste Hindu kid?

AMIR: Uh, I don't know. Maybe not, I would think not, because the majority of people would think if I were a Hindu kid, because you know the majority of them are Hindu, even though one kid he was Hindu so he didn't get as severely punished as I was going to get.

JONAH: Were those other kids who were getting punished, did they run away with you?

AMIR: Ah, to be honest, when I ran away I ran away by myself but somehow, like, I was the first one who ran away and later they ran away, so I kind of got name for that, that I made them run away, so I got more of a bad name.

JONAH: Okay, so, and when your family saw you on the roof tied up, did they, I mean do you think that they, once I hear it, it sounds like they would have supported you leaving, right, for your safety?

AMIR: I mean yes they would, they couldn't really come to the roof I was on, I was on a Hindu family's roof, I mean the people who tied me up they were like ... big people in the community, they have names in the community, they were top-level people, like those people who run small villages.

It thus emerges that Amir's departure situation is indeed both deeply personal and structural—it ties into family, and communal identity, and ridicule and shame and punishment.

Sachdev

Sachdev's remarkable story illustrates vividly the murky space, the gray zone, between the "typical" here (if it exists) and the idiosyncratic, shedding light on how children's journeys resist classification into neat and ordered taxa. Sathi's records of thousands upon thousands of children reveal something about such taxa, and also about how distinctive each departure can be; they show the patterns one expects and sees everywhere else—moderate but not extreme poverty paired with abuse and punishment, overwhelmingly manifest in departure by boys. But in Sathi's records, the idiosyncratic appears as well: "the peer group [was] influenced by the films & glamorous city," says one; "habituated with movies and left home casually," says another. One child was depressed by the trials of dia-

betes and thus suicidal. Another grew sad because his father sold his beloved pet ox, and, in protest, left. A different (internal, unpublished) document from the same organization lists the following among a range of other
unique "reasons" for running away:

> He had played some card games on the computer and was
>   caught red-handed by the teacher
> Father had made fun of him, for his failure [at] SSLC [Sec-
>   ondary School Leaving Certificate], in front of his other
>   relatives
> He was not interested in learning the trade of vegetable selling
> Scolded by sister for petty reasons
> He was repeatedly criticized for getting third rank and when his
>   aunt threatened to send him to work
> Mother scolded him as he was singing some mischievous song
>   to a girl from his village, which prompted the girl to slap
>   him.
> Beaten by his father for refusing to work in the coal mines
> While playing, the stone that he threw hit a friend who started
>   bleeding
> Had a fight with his classmate and he broke his head uninten-
>   tionally
> Ran away & married a boy in temple, later he cheated her
> Farmers' crops got spoiled due to fake pesticides [that he'd
>   been supplied with by the distributor while working in
>   a store]
> Father was assassinated by a group of factionists, in front of his
>   own eyes, he was very afraid that they might kill him
> He used to get bored sitting at home
> Wanted to give up worldly pleasures of life and go to the Hima-
>   layas for meditation
> She had heard that she can earn lot of money in Mumbai
> Left home in search of his mother
> Dancing in the Ganesh Chaturthy festival & had got late and
>   thought that he would be beaten by his mother

Parents abusing for being engaged in a love affair

Mother was a drunkard and didn't care for him

Father is a drunkard and tortures the family daily

Many cases show flight from home as either punishment for some relatively minor infraction—"scolded by his father & stepmother for stealing Rs. 50"—or *fear* thereof: "damaged a bicycle," "did not take the sheep to graze," "fear of scolding by father for taking a bath in the river." These cases may obscure a deeper unhappiness and a more constant reality of sanction. And in general, most cases manifest again the same core set of features: stress, poverty, abuse. Sometimes alcoholism, sometimes stepparents. Often, fear of shame or shaming—an emotion or experience with special gravity in such settings. One remarks, scrutinizing the microcosmic in such stories, that sometimes a trivial departure leads to a disproportionately deep, hard, and long exile, such that the departure itself, rather than the purported reason for the departure, becomes the seminal adversity to which the conditions of life will now respond.

Recall that by Sachdev's reckoning some twenty-five children fled Haspura's homes around the same time as him; Sachdev was around ten when he left; he was twenty-one when he told his tale. To that point, he had never managed to return home for good. On why he left, and what happened, he explained things like this:

> I was enrolled in many schools, but I didn't go; instead I used
> to play with my friends, who were from nearby villages. Be-
> cause I didn't study, my parents used to beat me up. One day I
> got beat up really bad; I got angry and left the house. I took Rs.
> 1,000 with me. I went to the railway station and took the train to
> Delhi. It was my first time there. I had my *bua*'s [paternal aunt,
> but used relatively broadly] phone number and address. I don't
> know how to read and write, but somehow I managed to reach
> her. I had no money, and I really needed it. So, I started work-
> ing in a plastic company. I worked there for about a year. But at
> bua's house, they used to take all my salary. I was not left with
> anything at the end of the month. Soon, a law was passed and

the company had to move to a different place. So I lost my job.
And then bua started bothering me for money. One day I got
really upset and left her house and came to the station.

One might wonder if going to a relative's house was "running away": this
bua was apparently not true kin, but kin in name and social relation only.
According to Sachdev, she was "not my real bua. She is my neighbor's bua.
So, my family had no idea that I stayed at bua's place for a year." Even so,
one might wonder how this auntie's sense of obligation to Sachdev could
trump any sense of debt to his parents back home; it must be that she felt
more beholden to whatever nephew of hers sent Sachdev her way than to
social relations within the village, or a cultural belief that a child must or
should stay with his parents. In any case, at this point, still early in his journey, Sachdev left this fictive aunt for New Delhi Railway Station,

> I stayed hungry for two days. Then the third day, I befriended
> the vagabond kids on the station and they showed me Hanu-
> man Temple close by, so I got food [there]. Sometimes I used to
> be ashamed of begging for food. But then the other kids would
> bring food for me. I used to work on a per day basis. Doing dif-
> ferent jobs, like making tents for an event. I would get money
> every night and would sleep on the platform. Then the kids
> would steal money [from me]. This happened every day and I
> got fed up of this.

Many recurring, "typical" features appear in this part of the narrative—
the relatively common post-abuse departure, entangled with school expectations; the secondary running-away from a first destination; the gradual
incorporation into a group of railway kids; the identification of religious
space as a site for aid; the quick embroilment in opportunistic and probably
risky child labor (opportunistic from multiple perspectives); the inability to
accumulate; and the tertiary departure from the other children's predations.
    It is around this point—leaving the other children—that Sachdev's
story, as he moves into a second kind of sanctified space, grows more unusual.

Soon, I found a *baba* (a guru). I told him my problems — current problems, not the ones related to home. He took me to Hari Ashram in Kamla market. I stayed in that ashram for one year. I got food there and sheltered, and I only had to serve God. There are many other kids like me there. Sometimes people visiting the temple gave me money. I used to give that money to whoever needed it more. After staying there for one year, I left that ashram. I wanted to work. Then I found a man, he offered me work and took me to Ghaziabad [an exurb of Delhi]. He welcomed me for two days and the third day he started asking me to do household chores and all sorts of other work in the fields. I did not want to do all that. But I did, for the sake of money.

Note here the second instance of inter-child altruistic giving, of money in this instance; and a relatively typical trajectory again into informal child labor.

After another year, I got money — Rs. 3,000 — and then I thought: it is time to go home. At home, some were happy to see me, but others were interrogating me and asking me not to not run away [again]. I got bored there, and I wanted to work. So, I ran away again. To the ashram in Delhi. The *babas* were happy that I came. There was this one *baba* I had to serve and take care of. I did that, and sometimes I used to get a little bit of money.

By my count this is Sachdev's sixth consecutive departure from one site or another since his initial flight. "I had to do it," he said elsewhere, about the necessity of this continual running-away, "no other option." In certain ways it shows an interesting interface between exploitation and autonomy; Sachdev serially exercises his ability to produce change by controlling the one thing over which he has some autonomy: where his body actually is, a self-space nexus. In the absence of all capital, mobility is his only resistance.

As he grows, Sachdev's story grows ever-more distinctive, and diverges from any typical long-term outcome we might expect (even as it intersects in interesting ways with Aravind Adiga's (2008) Balram Halwai in *White Tiger*): not long after this second stint in the Ashram, he says:

> I met a politician [Congress Party], named [X]. He talked
> so much about his huge computer shop. He said he will give
> me work and said I would earn Rs 7,000 per month. So he took
> me to Gopalganj. His house seemed like a *home*. I really liked
> this one person in the house. I used to call her bhabhi [fraternal
> sister in law, but used widely]. She taught me computer skills.
> Then I used to clean their car and all that, and soon I learnt
> how to drive. But they never let me get a driver's license. They
> thought I would run away, if I got independent. After two years,
> at Holi everyone was going home. So, I said I also want to go
> home. He [presumably politician X] was going to Bihar, so I
> tagged along with him until then. There, he gave me Rs. 1,500.
> That was all my two years of earning. I was expecting my earn-
> ings to be much more. I was angry but had to accept it. Then
> I went home, celebrated holi, and ran away again in two days.
> Okay, so after that, the guy started calling me again and wanted
> me to come back to him. So I went to him again. But he didn't
> do anything that he promised me, like getting the driver's li-
> cense, more salary, and so on. So I was pretty upset at this situa-
> tion and I left working at their house. I had nowhere to go and
> no money, and soon I met another person. He was Lalu Yadav's
> brother in law [Lalu Yadav was then Chief Minister of Bihar].
> He offered me work for Rs. 3,000 per month including food
> and shelter. I stayed there for six months, driving for them. But
> they never gave me money. I used to ask every day, but they kept
> postponing things. I wanted to leave but they used to threaten
> me if I leave. One day, the brother was drunk. He took me to a
> construction site. There we got bikes to ride. He pushed me and
> I fell and broke my bones. I was admitted to the hospital. When
> I recovered, I started driving for them again. Then I asked for
> money and they started fighting with me. They said a lot was
> spent on your medical expenses. I said the medical expenses

were Rs. 3,500. So give me the rest of my money. He didn't give
me anything, beat me up, and then I left.

Now here I am. I am getting Rs. 3,000 per month here
plus Rs. 50 per day.

These days, Sachdev remains a driver in Patna, the capital of Bihar. If
we assume that the Rs. 50 per day is a living allowance, he probably makes
just enough money to survive but not to thrive (Bihar has one of the lowest
GDPs in India; Sachdev was at the time of this interview making about a
third of what is India's 2016 per capita nominal income, and less than half
of Bihar's; if we factor in the Rs. 50 per day, his income still comes in more
than $130 below Bihar's for 2017–18, and at exactly half of India's for 2016;
while not exact science, this gives a sense of what this income might mean).
Though he visits the village (he'd visited around a year-and-a-half before
the interview here), it does not appear that he will ever again consider it
home, nor that it ever could have been, at any point after his departure.

Above all, as I will show below, it is Sachdev's *assessment* of where
all this running got him—which is to say, in his view, nowhere—that is of
the greatest interest. Unlike the returnees who go back home to feel celebrated—heroes with an epic journey behind them, now able to fix everything—Sachdev felt himself a failure and could neither not stay home nor
rectify the miseries with which his departure surely shares a common origin.

Dissimulation

Of course, it is not always possible to say, for sure, what the runaway's true
story is, what really happened. In such cases, the fluidity of the story itself,
and the manner in which it is fluid, becomes a question of interest and an
object worthy of its own consideration, and decenters our focus from the
typical situation to the typical story, and from the story itself to what we can
learn from form and narrative, from the complexity of the telling itself. That
too reveals a great deal. It might, for one, illuminate how a child might be
drawing upon and engaging existing cultural images, models, and proto-
types of running away. Take, for example, Firdaus, who originally reported
to the NGO he was living in that he had run away from a very rich family

in the city of Agra when he was found by an NGO, a different one. When
I met Firdaus, he was considered a kind of "star NGO kid," exemplary, a
success story. I found him well composed, well dressed, confident, atten-
tive, and healthy looking. After I left Delhi, however, some more complicated intimations emerged. Khushboo was able to piece together and observe more of his story over time, or at least more narrative, and indeed it is
the *story* that is the object of observation here. In the tale Firdaus originally
told, the imaginary rich-kid version, when he ran away, some ten cars were
searching for him and ultimately succeeded in collecting him and taking
him home. He then ran away again, he explained, and ended up in the second NGO, where he caused trouble. One day Firdaus received a call that
his mother was there, who then got told the story that he had been retaining
his connection to his paternal grandmother, something his mother did not
approve of, as she was allegedly mistreated by his father's part of the family,
all of which resulted in a dispute. The mother was, it turned out, very poor;
his father and stepmother were, likewise, wage/day laborers in Sirsa, a small
city in western Haryana not terribly far from the Pakistan border. The initial tale reads, contrasting as it does to the reality, a bit like a film—with
hints of Bollywood action—which could potentially suggest a longing for
something else, a retelling that permits a distancing from the bleak, indigent
truth by bestowing on it heroism and high adventure.

Firdaus's dissimulative story of high origins may also be tied to his
effort to distance himself from other street children and to posit himself as
superior. Firdaus expressed very well the curious but common ways that
runaways have—and situate themselves within—their own moral hierar-
chies, in which certain street children's behavior (theft, addiction, gambling) is labeled degenerate by *other* street children. Solvent sniffing is paramount among indices of such expressions of superiority. Consider Firdaus
on this, who says that he is decidedly not a *sadak chaap* (street child, formally speaking), because, in his words, "I live differently. I don't speak to
any of the drug addicts." "K" is Khushboo here.

> FIRDAUS: I feel like going home sometimes but now I like
> staying in Delhi only and nowhere else.

K: On the streets of Delhi.

FIRDAUS: I don't live on the streets!

K: Where do you stay then? You live on the pavements of Meena Bazar.

FIRDAUS: (getting angry, raising the tone of his voice and speaking very fast) I don't live like other vagabond kids (*is tarah to nahin rehta jaise aur kanglē* [this is the name that gives the book its title, *kangaal/kanglē*, bone children] *rehte hain*). They are always high on solution. Have you ever seen me sniffing one? I also don't chew tobacco. I used to smoke cigarettes, before, but not anymore.

Does it matter that Firdaus's story differs so radically from what comes out later? It is telling that Firdaus doesn't seem to think so. When a disparity is highlighted, he either patches it by explaining why it might have been true, in some sense, or seems not at all ashamed. What is certain is that the *story* matters, that Firdaus—and Akhil, to boot, whom I discuss elsewhere—draw from a lexicon, an amalgamated corpus, of what he believes to be legible, accessible, credible street-kid stories, and sometimes even fabular, *un*believable, interesting stories, and that out of that bricolage emerges a kind of pastiche that itself reveals a composite truth both of narratives and of blueprint-like perceptions of what is important about the runaway experience and narratives thereof, and also of evasion and the imperative of dissimulation.

This is worth dwelling upon, at least for a short while, even if it forms a digression from the question of the child's journey from village (or city), of origins and trajectories, for it bears directly on that question. Dissimulation is a big deal in this process. The stories in which dissimulation occurs are where the information we have on departure and journey comes from, or does not. They are also the very thing, the very process or ritual, that is an autonomous, intentional, agentive effort to obfuscate origins, which should tell us something about the *role* and stature of origins for the child in the sort of situation we are talking about. Such storytelling, by design, obscures the pieces of identity that are rooted in origins—in place, in kin, in former institutional iterations of self. For Indian child runaways, those origins are, in general, to be negated or refashioned. The origins are the locus of various forms of power, various intimate tyrannies, which the children have identified as oppressive and unfree and thus from which they have elected (usually elected) to run. A kind of narrative hiding is possible, or a hiding through narrative, and the irreducible, elemental kernels of that narrative are the personal name or the natal place. Child runaways in North India engage names, places, and old selves in exceptionally creative and complex ways to obscure former iterations of self. It is as though the street child, as runaway, is a new incarnation, an *avatara* with only a tenuous relationship to its past existences.[4] Renaming and re-placing has the added effect of recalibrating or even reinventing the child's relationship to communal identity, to caste, to a self that came before. I am not, it asserts, who I was, nor do I need to be.

Firdaus and Parvati

But Firdaus's journey, if we pick it back up where I just left it, has more to teach us than just what might be read from the fact of its loose relationship to actual event, from its dissimulative, narrative, or literary features. A much more profoundly complicated picture resolves itself in subsequent developments and conversations that suggest something about the real nature of his departure.

When Firdaus's mother came and took him home from the shelter, she was upset to see him living like that: "dark," as Khushboo described the mother's characterization, "skeletonish, shabby"—and she "cried and cried like anything" in front of Khushboo's house, pleading: *what do I do?*

Most stirring, poignant, and unsettling of the materials we have from Firdaus come from an interview (part of the ongoing NSF-funded research for this project) between Khushboo and his mother—in his presence—from Delhi in April 2011, the same season I was there, a few months after I first met him. Khushboo had at that time already had a protracted, intensive interaction with Firdaus, partially mediated through NGOs. The interview illuminates many of the core features and themes associated with running away that are described throughout this book. I include a lengthy segment here, annotating Firdaus's mother as Parvati (not her real name—but her real name is a Hindu name, and Firdaus's real name is indeed a Muslim name). Khushboo is again "K." I have elected to conceal NGO names here as well.

> PARVATI: He ran away because of the complaints by the world that his mother is like this or like that. Doesn't his mother love him a lot? Even now he tells me, "Mother, don't speak nonsense" [*Ma, tum faltu na bol jiyo*]. I have no brains. Only my son has all the intelligence. But if you [Firdaus] are so intelligent, then why do you keep leaving me and running away?

> K: We have used all possible means of love and anger with Firdaus but he just refuses to do anything.

> PARVATI: Yeah, I heard from staff at Center X, who had high praise for him, saying he is very intelligent. Even if I agree that I used to beat him up, force him to work because of which he could not study, now over here he is free from all those compulsions [*yahan to puri choot hai isko*], free from the jail he feels his home is, then at least he should have listened to *you* and worked according to his will. These people don't beat you. I left young children at home and I have come here. Hungry and thirsty, I have spent two days with you here, but I fail to understand why you [Firdaus] are staying in an environment like Jama Masjid [the Congregational Mosque, an informal center for street-dwelling youth and homeless folk]. You should have rather stayed at this Madam's place [pointing at Khushboo] and done their dishes and other household chores, that way they would have helped me as well. I am a poor woman; I don't have any job.

> K: What work do you do?

> PARVATI: Both my husband and I work as daily wage laborers [*dihadi mazdoori*] in Sirsa.

K: Firdaus told me his father is a big property dealer and a fleet of cars come to get him.

PARVATI: I don't need to lie. Yes, I have a huge family but I don't consider them much as they got me married in a poor family. They might have given me birth but did not do anything for me [*janam diye karam na kiye*]. But I do what I need to do. I live a respectable life, I have never been disrespected.

K: Why is Firdaus like this then?

PARVATI: His father also was like him. I was fourteen when I got married and Firdaus was born when I was about fifteen, sixteen years of age. Soon after, his father passed away. After about two years, I got remarried. My husband never considered him a stepchild [*usne kabhi ghair na mana*]. But Firdaus — under others' influence — has been away from home for three years now. The first time, I looked for him for fourteen months and finally found him and took him back with me. My in-laws in Meerut had ill-treated me — a lot. Used to beat me up and not even give me enough food. This boy, he went and stayed with such enemies of mine. Working as a daily wage laborer, I made a house. My mother-in-law used to speak ill of me all the time. I still stayed with them and provided for them. Finally, with three little children, I left that house when I couldn't bear it any more. This boy felt no shame in going and staying in that house. His father provides everything for him, from clothes to food to medicines, but he [Firdaus] still blames him for being a step-father. Now if Firdaus tries to incite me against him, I won't listen to such nonsense. I don't understand what is he getting out of staying on the streets.

Had he stayed with you [Khushboo], even mopped your floor, you would have considered him a good child. Instead, he takes me around Jama Masjid and calls such dirty fellows his friends and introduces me as his mother. I would want to turn my back at them. I put him in a madrasa, he did not settle there. So I brought him home and put him in school. One day, I raised my hand on him, so he ran away from home. But when I took him back after fourteen months, I never beat him up as I was always scared he would run away again. I let him stay at my parents' place. My father is a very popular mason [*thekedaar mistry*] who is in great demand for his talents amongst rich people. He would try to teach Firdaus work and my mother pampered him. He still did not learn anything there and roamed around aimlessly in bad company. My elder brother has two cars. He once drove from Kolkata to Delhi. His son also owns a car and drives to make a living.

I looked for Firdaus all over when he ran away again. I lost my mobile too and my parents have been so worried about me. Every time I went to Meerut to visit my sister, I first looked for him in Delhi. I went to all possible saints and sages. So tired, I stopped trusting God. I blamed them for first taking my husband and then my son like this. Finally, I went to your Hindu gods and prayed before Lord Hanuman. I told him if you have any power, get me the whereabouts of my son, even if he is dead or alive. I would cry and go crazy, his father would take care of me. When I went crazy, I even broke a television and CD player. I have broken two television sets since he left. Have broken so many things. I feel: why I am keeping those things if my elder son is not with me? I went to one *maulvi* [Muslim cleric, or possibly, in this case, holy man] who said Firdaus was working in a factory and will return in a few days. When Firdaus did not return even after eight days, I called the maulvi again. But then fifteen days passed, then twenty, then twenty-five. I felt the maulvi was a dog. Then one day I was sitting somewhere when a Hindu priest passed by. He could sense me crying inside for my son. He asked to get a clove, and after some recitations, said Firdaus was in Delhi. I told him I won't trust him until he gets my son back. I was so sick crying all the time that I was weak and bedridden, then. Three days later Firdaus called on the neighbors' number. This was two months ago. He had been away for two and a half years by then. This was his first call in as much

time. Though he visited Meerut and met my husband's family,
they never told me and I also don't go to their village. His own
grandmother, my first husband's mother, has been crying for
him. She is so loving and caring and always took care of me
even when my husband passed away.

My husband and I had gone for work when Firdaus beat
up my younger son, took his mobile and left. He went to a
neighbor's house and told them he had beaten up his younger
brother, and that his mother would beat him for this. They said
your mother doesn't love you. If I did not love my son, why
would have I raised my child and taken such care of him? I
would have given him away and anybody would have taken a
male child. But he was the first fruit [*pehla phal hai*].

What emerges here, in this remarkable maternal narrative and through Firdaus's many—clearly self-protective and fancifully if self-fashioningly com-
pensatory—versions of self and story, is a perfect storm of the elements
we've already seen: labor and strain, complex configurations of punishment
and physical harm in many directions (and of various people accusing each
other of the same), death and loss, feud and vengeance, and a mother's
sadness. Friction with a stepfather, attributions of bad-seedhood and no-
goodhood, and regrets and questions over love itself. In Parvati's calcula-
tion, labor—informal labor, in this case—is associated with morality, back-
bone, and upstandingness. In Firdaus's, or at least in Parvati's projection of
Firdaus's, labor is prison. Among other interesting elements of the narrative
here are the fluidity of religious affiliation and exhortation and the reference to just how easy it might be—and how normalized—to give away—
or possibly even sell—a child, as perhaps in times past. The conversation
continues:

к: What does Firdaus want to do now?

PARVATI: He wants to return home, but I cannot understand
why he doesn't stay at home. I took him back earlier too, after
fourteen months, now I am taking him back again from you.

I am worried about the kind of people he is hanging around with here. He brags about places he has visited—Pune, Bombay, but he doesn't notice he missed out on his mother's care in doing all this [*ghūma par maa ka anchal to na dekha, ki maa ka anchal kitna pyara hai*]. My husband too takes care of all the five children like a mother. Firdaus is so big but he took care of him like a child too, and of all my four sons including Firdaus and two daughters.

K: When he went home this time, he said he will stay for good and not return to the streets.

PARVATI: Firdaus said, "Khushboo didi has paid my fees of Rs 12,000 in the cell phone-repair course and two months are left in finishing it." I said you have stayed away for over two years, so let it be another two months and finish the course if it does you some good.

K: But he never went for it after he returned.

FIRDAUS: No, I did! I went for six or seven days but then I stopped. Every time I call home, she starts pestering me to return home. They go to all kinds of *baba* for my return.[5] They are driving me crazy. You both are just finding fault in me here! I had to stop going to the center after I returned. I returned to Delhi to see things.

K: See things? What are you looking for in Delhi? Please make me understand.

PARVATI: Every time I call, he says he is at India Gate. My younger daughter is so fond of him since he returned and wants to talk to him all the time. Don't I understand the kind of atmosphere he lives in here—bad area and bad company! Firdaus was talking to someone and I overheard that he failed in the cell phone course he was doing. Even when he came home, he was so rude to his siblings. Never spoke politely or with love. Always bossed others around. All my other children are so good in their studies. I never have to ask them to study.

K : Is he coming with you to Sirsa right now?

PARVATI : Yes, he says so.

K (sarcastically) to Firdaus: When do you plan to run away and
return to Delhi again?

PARVATI : Well, if I say anything untoward [*agar main kucchh
bolūngi*], he will immediately run away. Since he has come from
home, he has become so weak.

My husband earns, we have a small family. During season,
sometimes I work or else sit at home.

K : What do your neighbors say?

PARVATI : My neighbors say all sorts of wrongheaded things
[*ulti-seedhi bolte hain*]. This boy runs away hearing all that.
Women say all kind of things about his running away, that he
is *harām ka beta*.[6] If he was *halāl ka beta* he would have stayed
with you.[7] Especially during fights, they [the other women] at-
tack me, saying if your son was true a son [*asli jana hota*], then
he would have stayed with you. I hear all kind of things because
of this boy and he still says he will beat up Bhagat.

K : Who is Bhagat?

PARVATI : A baba. He had come recently. I went to a lot of
people but did not get any help. Wonder how this man . . .
seems like God sent him to my house. He himself came to my
house and spoke to me. I told him about Firdaus. He must have
done something such that Firdaus returned after two or two
and a half years.

I have told Firdaus that if you run away or leave, then
prove yourself [*insaan ban kar mujhe dikhana*, that is, show
me after you've become a (full-fledged) human]. But this boy
has run away twice—and look at his situation. There are three
or four Bihari women—wives of alcoholics—in our locality
who are bad.[8] They brainwash Firdaus, tell him "your mother
and stepfather are going to make you work and take your earn-

ings." If both parents are working, if he helps with some household chores, washes dishes, what's the big deal about that? If he filled water in buckets, he worked in his own house, not anyone else's — "my parents are out helping, so let me help." His younger brother also helps with household chores despite going to school. Firdaus is even older and he should help. If he ran away from home, had he proved himself by doing some good, at least my neighbors would not say all kinds of things to me. I have never wanted to eat from my children's earnings. He is right here. Ask *him* if he has ever earned and given me a penny from it. My other boys are earning after school. I take their money and put it into their own account, so it can help to better their future. I have my interest there. I had Rs 40,000 saved once and I gave it to a person to make gold jewellery for my daughters for their future but I was duped by the motherfucker. Now I put money in my children's accounts. During season time, we scavenge from the waste, recycle and then sell it. Sometimes we make Rs 10,000 or 20,000 a month and can incur a loss too sometimes depending on inflation (*mehengai aur saste ke upar*). Two years ago, when he had run away, I suffered huge losses. I had lost all my money. I am investing again this year.

"They left after a while," wrote Khushboo. "Parvati cried again when I came to see them off at the gate."

A poignant and distressingly enlightening story — still more interesting indeed than it might have been by virtue of its triadic structure, with Khushboo, Parvati, and Firdaus all there. Even through Firdaus's general silence, positions and positionings are revealed by the directionality of address, and content that varies by virtue of who it is that is being addressed. Of note in this latter portion of the interview is the complex narrative — indeed, the complex equation, or configuration — of shame, of bad-kid/good-kid narratives, of embarrassment in others' eyes, and above all, of the idea that a runaway might need, in the wake of an urban sojourn, to earn back his parents' moral approbation through earning *money*. What that intimates, of course, is an equivalency between morality and money where the trespass is forgiven, on some level, by a peculiar form of repayment (no questions asked, usually, about *just how*). An emotional debt that translates into fiscal value. It is captivating, moreover, to see the flexibly religious terms in which these expressions of sanction and worth and repair are couched.

Firdaus at length and in time capitulated, went home, began, in the vein of street-child, class-based "reform" that I mention elsewhere, computer courses, and he enacted lifestyle changes and donned "proper clothes," and in the meantime gained middle-class friends from his courses and shunned his prior street life.

But then, as might have been predicted, he left his course in medias res and began, entering the proverbial "informal economy" again, to sell vegetables, to work in construction sites, to work as an assistant on long-haul trucks. Firdaus's mother was distraught.

Firdaus, for his part, reported that his mother suffered burns while cooking and that he was in Delhi expressly to get money for the mother's treatment, to the tune of Rs 20,000. But then, while he was back in Delhi, he reported that he had received a call that the mother was dead, that she had killed herself by self-immolation.

Of Sadness and Love

I try not to impose too much of my own value on all this, but I do want to ask whether all this is fundamentally *sad*. That is subjective and it is not entirely or particularly measurable, but it is, by virtue of being human, legible. Other people's sadness as reported by them is a measure in and of itself. It would be nice to say, well, maybe this is all just *our* culture speaking, maybe we don't know, who are we to say, maybe this is just a different model of living, perfectly normal. A different lifeway rendered legible through cultural relativism. But no: that is not how it feels to me. Not entirely. I spent my seasons in India working on this, as did Khushboo, filled with a tremendous certainty of these children's lament, their shame at leaving their mothers or siblings, their sense of wrongdoing, their memory of deaths precipitating departure, their bitterness at parental divorces, scores left unsettled, vengeances unconsummated, laments unexpressed. We saw a lot of tears. Amir reflects:

> For me the reason for running away, I think . . . they alienated me, they gave me so many different bad labels, and you know, I got beaten many times and I hate it, I hate it, I did not like it all getting beaten by people, family, just you know in general and also I wasn't really treated well in my home, so that was the reason, I was very emotionally driven to not go back, and when I think about it basically I was just on my own. I wasn't getting any emotional support, any kind of support . . . so many kids run away because they don't have any resources, they're abandoned, they're harassed, they're thrown away, they're missing families . . . there are some kids who are just like, nobody just loves them, nobody cares for them, they leave them alone, they don't care if you're going to be dead or alive, they just treat them as an object.

And thus sadness, and love, really do figure in. And there in the city, they occupy drab wastes they hate, face daily rape as tax, find themselves trapped, confined, shackled, and imprisoned, get the little money they make stolen, end solvent addicted, their brains eaten away, sick, and homesick, without a way back home. Too many die. Many live and wish for death. The scars they inflict quite universally on their own skin bear testament to all this.

The Remembered Village: A Home Deferred

What do these children become in the city, over time, and what does the past become to them? How does the village ferment to the point of forgetting as they range new terrains? There is a sense in which they cannot any longer be said to be village children stuck far from home. Indeed they are not. They are a species unto themselves, a certain type of city child, in fact, who is perhaps more part of the urban landscape than the actual city-

Down (and out) in Delhi

born child. They do not evoke to passersby the villages they've left. They
are part of the fabric of intersections and stations, of scrubland and park,
shrine and alley, not countryside. They inhabit the city's actual physical
spaces more than the rest, who but pass through them or use them. Even
in runaway children's peer groups in the city, the village is of little consequence and rarely named. Biographies are recounted rarely and sparingly.
Nobody cares where you came from. You're here now, and that's the fact
of it. And yet the village figures prominently in the child's internal narratives of self, trajectory. It is never forgotten. Sometimes it figures into an
intended future itinerary, a return. Sometimes it figures in as an itinerary to
avoid, an I-will-never-go-back-there. Some go back swiftly, after only days
or months. Some go back many, many years later, so long sometimes that
parents are dead or that the departed child was believed dead. And some
indeed never go back. But always the village—even when it is lost as a piece
of the child's visible, publicly accessible identity—is there as an absence or
presence, a distance or proximity of varying degree.

Is it as though the self, in its removal from place, is in part disaggregated, disconnected from some of its components of kin and memory, as though home is a ghost following the runaway around, haunting or possessing, from varying distances. It can be exorcised, but only in part. That chthonic part of the person is now removed physically and conceptually, left far behind in both time and space, to be filled in with new material from a new place, while meanwhile the cord of connection between child and home, that natal umbilicus, is stretched to its limit. But the memory cannot be shaken. The village is like a root, or a mark, or a scar—something fundamental about the person. Village enters the domain of memory. And just as the village is itself at times *reduced* to a memory, a distant thought, at other times it *overtakes*, consumes the self, comes too close for comfort, becomes everything about you: what you are, what you want, what you aren't, what you don't want. I have marveled at what form the village takes, distilled into memory and fluid, formless possibility like that.

But what role does that memory play? In naming this subsection I invoke the classic work of Mysore Srinivas—*The Remembered Village* (1976)—not by accident and not entirely lightly or for the sake of pun alone, but also to draw a connection. When the lifework of Professor Srinivas, the intensive ethnography of a village, was burned in an arsonist's fire, he was crushed but endeavored, at length, to reconstruct what he had observed in years of ethnography by memory alone. To the runaway, distance in space and time means the texture of life in the village takes on, in narrative, a different varnish, gains fluidity and poetics, is deployed and mobilized in the mounting of campaigns of self-protective deception, is variably idealized and shunned, becomes a fiction of sorts, reliant more on memory and its uses than on relationships and quotidian rhythms.

"What," I asked Amir, "do you remember from the years when you were still settled . . . when you hadn't left? Like how would you describe it to someone?" He replied:

> Okay so what I remember, my village . . . it was a village where
> you had no—like I didn't really know about the streets, the
> trains, the cars, and all . . . you know, where my family used to
> live, which was there by the river, where we used to grow crops,

live there, go to the fields, play in the village, play marbles, pick
things from here and there, go to the well, take out water, take a
shower, go to the river, poop there, take a shower there, I mean,
that's all I remember, when I was that young.

In the transit to the city—the thing to which home is eternally in a binary
relationship, the thing that is most fundamentally not it, that is rivers and
fields to its streets and trains and cars—and in the discursive genesis of
village-as-memory, home becomes a kind of impossible place: impossible
to return to, impossible to have as it was, impossible to feel the same way
about, impossible to be part of, impossible to stay yoked to. The runaway
cannot, after all, ever be the same child he was back there. The passage
not only of space but of time stands in the way. Indeed, it is impossible to
bridge the gulf between the narrated and real homes.

And yet, in narratives of and constructions of the departed village, I
have also seen that home looms large as a domain of shame and betrayal,
above all betrayal by sons of mothers, and of an inceptive moment of departure that forges who one is now. It also thrusts the village into a new realm
of possibility, however, beyond memory alone—now the village can potentially be a range of things to the departed child, things back there are now
less determined, and new relationships of many forms can potentially be
forged. Home doesn't need to be what it was before. By leaving, the child
has opened it up to reauthoring.

There are two types of *longue durée,* of long range, in this book: first
the one that is Braudelian (Braudel 2009, for example), the long view of
history and time, what David Christian (2005) might call big history (or
Shyrock and Small (2011) "Deep History")—how the centuries come to
life and take form in the shape of a moment, of things, of predicaments.
Then there is the long range of a runaway's life, stretching away from the
present in both directions: memory, behind, and possibility, or the lack
thereof, ahead. Five years on, ten, even twenty, and some of these kids are
still there—in numbers—in the places they ran to. "Street" becomes habitat and home, while "home" recedes in memory, in the possibility of return.
For many, there is no return; shame and failure win out, the marks on the
body, the missing limbs, or the changed, no-longer-little and no-longer-

cute body; the shame of what had to be done and what was done; the sense of betrayal and dishonor; the unmentionable truths of violation, of why you left and how you got by; the illness slowly taking over; the addiction. Time snowballs on itself; at six months it already feels too late to return, to look your mother or brother in the eye. At five years it is terrible, irreversible, and profoundly painful. And over time it gets easier and easier to cruise along in the city the way you've been doing; you've mastered it, you've forgotten other places and their ways. It beckons, the city, it has a pace, and after all, this is where you're comfortable. People know you here. Why go home to the things you ran from? Why see everyone cry or turn from you in hurt or fail to recognize you? The village is the past, the city present, in both senses.

Not only do these long-term runaways sometimes stay in the city, however; they also often stay precisely where they arrived, or at least precisely where they ranged about as a street child. Sometimes they graduate to a kind of bosshood, in charge of a group of juniors, newer arrivals, whom they "manage" (and from whom they exact, as I have mentioned, all manner of tribute, sex, money, and labor alike); sometimes they are able to acquire a local stall or they are given control of a teahouse or stand they worked for before; sometimes they carry on begging. I saw this sort of long-term employment arrangement at my favorite Kali Mandir tea stall in 2015 — the same runaways and orphans I'd met in 2011 were still there, working for the same man in the same spot, only now they were themselves men. Filling the slot they'd occupied before was a cadre of new young arrivals. In most cases, those who stay on do not enter the mainstream labor markets, the larger systems of manufacture and production that draw workers from rural areas, but they do cement new connections of city and country and sometimes, reviving contact with home, the more typical relationships with villages that other migrants exemplify. And sometimes, of course, they go home. Amir did.

Return

But what is it that happens at that fraught moment of return? Must it always be fraught? It is important that it is often accompanied by shame, a shame at betrayal, at having caused grief and fear, at being accused of the same, at missing the life or a death of a loved one or being accused of having caused that death, at missing childhood altogether or a sibling's childhood, at having taken an easy way out of a hard situation, at having emotional debts to repay, and so on. Sometimes, a child may feel so afraid of whatever it was that was being escaped—abuse, above all—that he may feel that it never is the best time to return, or may await adulthood. Sometimes the return is a grand spectacle. Sometimes the child slips right back into life as it was, if not much time has passed. Sometimes the return is a kind of surrender: *I can't make it anymore in the city* or *I need my mother* or *I am dying* or *out of money* or, indeed, *running away from the city itself,* from a situation of victimization there that paralleled and bested (or worsted) the original reason for running, a situation that needs to be gotten out of. Amir ran away from a setting *in the city*—to which he was already a runaway—that he needed desperately to get out of:

> AMIR: When I was at the station, so one time . . . there was no
> food, right, so somehow I sneaked into Platform Number 5 and
> I was able to get some food but those kids they were so angry,
> and I knew that if I get caught, they will kill me. And like, you
> know, it was going to happen that [but] somehow I was able to
> get out of that railway station . . . I just got a feeling, so I got out
> I just, at night, I just somehow walked out of the station and I
> was scared of a ticket collector but nobody was there and I was
> just—I was able to sneak out.
>
> JONAH: Did you feel free?
>
> AMIR: I mean, I just, I feel I saved myself from getting beaten
> or hurt by those people.

It sounds like Amir's original departure from the village. And that, for its part, was merely Amir's flight from a relatively short-lived situation within the city (and if the exploitation was worse than the peril of violence, he does not say it), to show the will it takes to run, even there, and also the will that can be exerted. But in certain situations, the returnee who makes it *all the way* back home can also become a kind of celebrity with magical, esoteric knowledge—of the city. The returnee may himself feel that as well. Later Amir did this as well:

> JONAH: Did you think of going home right away? Did you feel guilty or worried?
>
> AMIR: After spending a few months, where we had to survive and then where we finally figured out, "man, life is rough here," every day figuring out where to sleep, every day figuring out where to find food, I mean, every day, you know, it's hard to survive, it was pretty difficult, so there was a point that where I decided to go back home, then there was so much anger in me, that really refrained me to go back home.
>
> JONAH: And did you go back home eventually, or not?
>
> AMIR: Uh, no, I went there, home, when I was settled down and I started going to school, studying, then finally a social worker, they said we should definitely go back home, and just try to see parents and just have good relationship with them and then come back, if you don't like it, come back.
>
> JONAH: Right, so what happened when you did that?
>
> AMIR: Oh, so, when we did that nobody really recognized me in my family or village, because I was, I was totally transformed, you know, in terms of height, weight, looking, and then my family and village thought that I was dead, and didn't really care but then, like you know, I have other siblings so they were very glad that I was still alive, doing fine, and got bigger, taller, and then just a lot of people, they came to my home, they sat down and asked so many different questions that sometimes I had no answer for the questions.
>
> JONAH: Like what kind of questions?
>
> AMIR: Like, you know, what did you do, how did you live, I mean like, how long the train looked, what kind of food you

ate, what is *chhole* [a type of chickpea dish], you know like those kind of questions, because the countryside has a lot of different names for dishes and different things, because they are not aware, so they don't really know much. So certain questions, what does school look like, what did you learn, how did you work, you know, like those kind of things.

JONAH: So you were almost like a celebrity, almost famous a little bit.

AMIR: Yes, exactly, yes.

And in this way, the from-the-city village returnee becomes a possessor of an otherworldly knowledge and experience, as if that person has voyaged to a netherworld, or space, and returned to tell the tale.

JONAH: How did you feel when you went back to the village? What did it feel like to see your parents and see the village?

AMIR: I mean, I felt good, but, I felt good but I wasn't really so happy about it, I just feel like I got reunited with my dad and mom and they are happy to see me and I'm happy to see them, like, the things I used to do in my village I went to those places and see, ah, it wasn't really very exciting, it wasn't very exciting, there wasn't something that, like, that motivated me to stay there. I felt good, I felt sort of happy to get united, but it wasn't like an overly exciting thing for me.

JONAH: Was it because you had seen the bright lights and big city and so on?

AMIR: The reason is that, I remember, even though I was doing good things, but, they labeled me as a runaway kid, you know, they labeled me as a bad boy. So they still would call me as the name as they used to call me, like, those things, I mean I changed a lot but they never changed their mentality, they think I'm the same person, I've grown taller, that's it.

JONAH: The last thing I want to ask you, well two things, are you in touch with your family now, or not really?

AMIR: Yeah, yeah I'm touch with family, yes. Of course I am, with them.

JONAH: What do they think about who you've become? I mean they must be, are they proud of you?

AMIR: Yes they're very proud of me and, they regret what they had done but, you know, yeah they're really proud of me because, whatever happened it happened, it took time to learn, you know I helped them out, so, they are very proud of me yes.

JONAH: And do they, do you guys ever talk about what happened or not really?

AMIR: Um . . . if they ask, then I will let them know, if they don't, then I don't.

JONAH: Yeah, no I mean about the past, the original act of running away.

AMIR: They, sometimes they will remind me, sometimes they remember and they talk about it, just one-on-one.

In this case, one can sense that this part of the past remains a central part of the present, but also a site still of pain, to be tended gingerly, handled with care, perhaps nurtured for a long time more until it resolves itself properly. Amir clearly does not want to speak directly to the "what happened" question at this point, which is before the interview where he describes the roof scene—there is a fair amount of murk here still. On the surface, it might seem like Amir's response suggests that he has *rectified* an error. Looking deeper, and at how central this element of his life has become *to* his life, it is possible to see that running away might itself have been the progenitor of his—and his parents'—pride. And it is nonetheless not clear what exactly it is that "they had done."

I followed up on this part of the story in the later interview and gained a still deeper perspective on Amir's return:

JONAH: What was it like the first time you went back home? Did you see those same people [who had threatened you before your return]?

AMIR: I mean I saw those same people but things were very changed, like I was still scared, I was still scared, like the memory I had, all the same memory, came alive, and even though my dad and mom they were much older, they were excited to see me, but I had the same kind of name, you know, people didn't forget what I had done. So people would call me *bhagola*. *Bhagola* means *oh he ran away, you know now he came back.* You know, it wasn't a great feeling.

JONAH: Is that term *bhagola* a well-known term in the village?

AMIR: Yeah, *bhagola* yeah.

JONAH: So you didn't know many kids who had run away before you or did know some? Or you had heard of some?

AMIR: No, I didn't know, I think I was the first one who ever ran away from my village, the whole village.

On the one hand, this contravened my supposition of running away by virtue of a transmission of a model among peers.

JONAH: But was there any sense in which you were proud, I mean you're a little bit famous [*mashhoor*] for doing it?

AMIR: Yeah, yeah, yeah, but like also just because of like me, a few kids also ran away, so I got more bad name for that because you know, like, one of them got, died. One of them got into drugs. Like none of them had very good life except for me, somehow I was able to pull my life to a good direction.

JONAH: Right, as if it was somehow your fault.

AMIR: Yeah like somehow they are like blaming me, because I was like the biggest example for them.

But that, on the other hand, shows how one individual's innovation in leaving *generates,* in turn, a new model for others.

> JONAH: And yet somehow right now it's almost like this thing has become part of everything about who you are, like you know, where you are right now is because, in some sense, because you did that.
>
> AMIR: Right.
>
> JONAH: So is there some sense in which you're, in which you feel that that was a central experience for you?
>
> AMIR: Yeah it was, it was I would say it was an important experience, at the same time it was not a make-me-feel-good experience.
>
> JONAH: How about your parents now, do you talk to them on the phone or Skype or whatever?
>
> AMIR: I talk to them, they are very proud. See, not, like my parents, my parents are proud, the whole village is proud, but in the sense that they know that I have done a lot of different things in my life, I've been to the village many times, I've done good things for them all, so like, I kind of like, you know, was able to erase those bad times in terms of being as a child, because I just told them I was a kid so I had no idea, like I wasn't really mature, I wasn't really going to school, I knew very little, they weren't helping me out so I just did what I felt right to do. Also, like now, when I go to home, like, everyone wants me to sit down, talk to them . . . it's a good thing, they see me as a hero now.

Amir's story is unusual and interesting, as are the circumstances and nature of his return. We have to assume it is uncommon but accept that it might be more common than we think. If nothing else, it shows the interplay of sweeping structural elements and idiosyncratic stories. Perhaps many runaways come home redeemed, renewed, with a reinvented self. Roshan from Muzzaffarpur, Bihar, who derived the idea to run away to begin with, like so many children, from another child, describes running away as an act that can potentially negate the problem that motivated it to begin with. When asked where he got the idea to leave, he said:

> I saw someone running away. His name is Sanjeev. He is back home now. We had a talk and he gave me his phone number, which I lost right after. Now that he is home, his parents don't stop him from doing anything. And his father stopped beating him as well.

Indeed, return may create new landscapes that allow the act of having run away to form the foundations for a new fabric of something besides the crisis that spawned it, something that (as a new context) can be responded to—a fabric even of life and subsistence itself. Running away becomes the baseline situation that now is the social fabric, rather than just a response to a fabric in crisis. So when Masud, the parking attendant in his late twenties at Hanuman Mandir who was stricken as a child with polio, recounts his runaway narrative (left at twelve, returned at twenty-four, and then back again to Delhi), he asserts (as mentioned): "I went back and ended my family's poverty." A similar and remarkable instance is of Nuruddin from Cooch Behar, northern West Bengal: I was working with him as an eleven-year-old in 2007 in the courtyards of the upscale shops at Vasant Kunj (Basant Lok or Priya Market) when one day, in his friends' words, he left—some said he was taken by the police and others that he'd left home. I'd seen him only a day or so before—wearing a mask, curiously enough, as he walked through the crowded market. Four years later, in 2011, I returned to the spot to look around. I saw a teenage boy with a younger boy lurking in an alley. Curious to see how tight the peer networks were here, and how long they might last over the years, and with a strange suspicion, I asked if he'd ever met a boy named Nuruddin. He stared at me a minute. I *am* Nuruddin, he then said, astonished, I'm back. He was taken aback, and recognized me. We talked for a time. He'd returned home, it turned out, to Bengal and advertised life on Delhi's streets sufficiently well that—now a migrant, not a runaway—he'd returned to the capital with his entire family. Running away was his inceptive and pioneering act of family advancement.

In the same Rasalpur of Dumra Block in Sitamarhi that had seen some fifty children run away, one former child, one who had left with other children, in the wake of still others before him whom his mother said he had "copied," walked back up that path and returned home a staggering *twenty-five years later*, as with Saroo in *Lion*, after he had married and had kids of his own. Other children had heard of this "village hero": one recently returned runaway child from the same village knew the elder returnee's story of paternal abuse and departure well (alongside that of another child who had come and gone in two months), reported that the long-gone runaway had come to "own some property" in the city, and summed the situation up as follows: "We thought he did well. He studied there, and earned money. And recently, my friend said he'll run away too if his mother beats him." This underscores the way that return and the circulation of runaway stories generates a semi-heroic "model" for circulation—a "cultural" form that is fundamentally structurally engendered—in the garb of stories, and potentially for eventual emulation. Mohan, a now-grown former runaway who left for Delhi at around fourteen in 1994, and stayed away for three-and-a-half years, came back and like Nuruddin later left again, not as a runaway but as a breadwinner: his family, he says in this narrative of meritorious self-advancement, "was very happy . . . I stayed with them for a month, then I went back to Delhi. I used to earn and send money home. After father's death, I permanently came home and started a business." Most of his kindred runaways, he explains, have likewise returned to the village to work.

But most never make it home, and most never get to be heroes. This is the reality of it. Some are enslaved, trafficked away, before their imagined trajectory of "running away" is realized, and have no means of escape. Indeed some, if not many, among those who go back, perceive themselves failures, without any heroism: "I learned that if I work hard enough, I will get rewarded. But I have worked so hard, and nothing." Instead of being fêted, the now-grown Sachdev is stigmatized. People at home, he says, "are very mean to me. They ask me, 'how have you, who have been gone so long, achieved nothing? No money.' But I didn't get anything at all, so how would I save money?" Some long to go home, but, as in *Lion*, or as with Krishna in *Salaam Bombay,* struggle to locate it: there could be a dozen villages of the same name across India, or a hundred, or a dozen in the same state or district; perhaps their dialectal pronunciation of a place name is difficult to understand, or perhaps they know only a neighborhood name. Consider how this remarkable (unedited) account, of a nine year old abducted from his village, from the unpublished case history archives of the Bangalore NGO Sathi, reveals some of the profound challenges of "finding home" in the context of children's limited cartographic literacy in a vast country, a problem that applies in equal measure to runaways as to the trafficked:

> Staff tried to trace the address of the child . . . The child could not give the detailed information of his house except saying that he belonged to Andola. Attempts were made to find the place with the help of Google. Andola place was found in three different states, i.e., Mayurban in Orissa; Gulbarga in Karnataka and in Pune, Maharashtra. With the help of police and missing complaints registered, enquiry was done in Pune and Gulbarga, but nothing found. The child was saddened after being repeatedly questioned about his family and started ignoring everyone. When the child turned twelve years he was transferred to Alipur GCH-I (Senior Boys home) [and camp] . . . During intervention by the camp teacher . . . after fifteen days of the camp, child said that he belonged to Chayamal in Andola and his father is a locksmith. The camp teacher asked the child if he is from Aligarh as it is famous for locksmiths, but the child could only recall that his house is near Jama Masjid [Congregational Mosque]. It was found that there is a mosque named as Jama Masjid in Aligarh. After taking the help of local people and police in Aligarh, we found a place called Shahjamal in Aligarh. The staff was confused that Chayamal (place mentioned by child) could be Shahjamal and decided to search the family in Shahjamal.

> We then requested one of our acquaintances, X . . . . to trace the address as he belongs to Aligarh. X enquired about the child's

caste which the child told as Ansari. X went to Ansari Mauhalla
[an eponymous neighborhood] in Shahjamal. Ho! He found
the family! [The child's] family was informed about him. They
were filled with joy that after four years their son is found . . .
The family also said that their son was abducted by an unknown
person while he was playing. They filed an F.I.R. at Delhi Gate
Police Station of Aligarh town in June 2009 but no response.
They also reported to TV channels and newspapers but failed.
At last they lost their hope of finding him. Fortunately they got
their child after more than four years on first July 2013.

Among those runaways who make it home, some leave again for the city,
double runaways with layered street lives. Some, as I have already explained,
are triple runaways, serial runaways; over and over again, like Sachdev, they
leave. Included in the case histories in Sathi's internal records and archives
are data on "number of attempts to run away." One child is listed as having
left home "countless times"; another, seven; still another, "more than ten
times." Repeated running away is very well attested in these case histories;
sometimes it is referred to as "the habit of running away." In a village near
Khusrupur, Bihar, a town east of Patna on the Ganges banks, Ajay's mother
explained why—and how many times—he had run away:

> There was a kid who had run away before Ajay. But that's not
> where he got the idea. Ajay had ran away before. This was his
> third attempt. About six years ago, his teacher beat him and he
> ran away to Buxar, Bihar. His father found him and brought
> him back.
>
> Another time, he just left home without telling anyone
> and came back after two days.
>
> He never tells us why he keeps running away.

Ajay himself, according to an NGO worker, was—despite this account of
teacherly abuse—facing a punitive father. Roshan from Muzaffarpur, for his
part, had tried to run away already by the time he succeeded:

This was my second time. I had already tried to run away once. I had gotten pretty far from Muzaffarpur. But when I was sleeping, a policeman came and took me to his house, where they got me cleaned up and took me home the next day. And then my mom told me not to run again. My mother is nice but my dad is crazy. I had a second and successful attempt within two or three months of the first. I gave it a lot of thought. I was not feeling good at home

And if some run away over and over and over again, more than a few of this precarious lot, close as they are already to the other side, to death, as they are, never make it out of childhood at all. Their departure is a fatal act. For these unfortunates, the many children who die, running away is irreversibly final, a definitive beginning to a terrible end.

History lives through these lives and deaths, through their larger arcs but also through quotidian moments in which stakes that seem trivial are in fact very high, and go very deep. The next chapter, which zooms out from this up-close scale, is a meditation on this notion (but not an equation or explanation of it) and its attendant questions — on how people "live history," how history might get translated to life itself, what might be the histories they live, and whether the people in whose lives such macrocosmic forces are incarnate know it. Of course, it is imperative to know that not only do people — even "powerless" people — live history, which itself sounds rather passive, but they also *make* it and narrate it (which are themselves separate acts), even on very small, intimate scales, like Amir's effect on others in his village, and even on those scales they generate consequence, entailment, and, like a stone in a pond, waves of resonance.

Of Crisis, Calamity, and Village Fabrics

The Imagined Runaway: Antecedents, Images, and Terms

I begin this chapter with the social life of historically valent ideas: how models, images, and notions, once they entrench themselves, can produce, conjure, or manifest action.

Recall the faded and worn keystone of the Calcutta Gate, all but forgotten in its back alley near Nabil's haunt, and marked with the year 1852.

Four thousand miles away and a year earlier, in the depths of the imperial archive at the British Library's India Office, I'd come across something else from 1852.

I was looking for evidence that such departures from the village might be part of a timeworn "tradition" or, alternatively, that they might represent a new reaction to rupture. Hard to find, such a thing. Imperial agents, officers, and archivists paid precious little attention to the movements of children. In British India, children were, as now, recognizable generally only insofar as they required discipline and inculcation. In the archive, they are represented largely as subjects for correction: in the reformatory (see S. Sen 2005 or Balagopalan 2014 for very nice accounts of the historical dimensions of such institutional existences).

At great length, I located in the India Office Archive a crumbling handwritten manuscript, over 160 years old: "Scenes in the Cities and Wilds of Hindostan," written (ca.) 1852 by Robert Hobbes, an imperial private serving as a clerk to various lieutenant generals of the Raj. It was a travelogue-style account, melodramatic, hyperbolic, and reflective, but it tells us something of how the image of a certain type, an urchin with a certain form of mobility, got constructed for Anglophone publics back home

and in other colonial posts, or reflected existing constructions. On one of
the barely legible pages, Hobbes writes, "Not infrequently even boys, animated by devotional or romantic feelings, set out on distant pilgrimages of
which their relatives know nothing till they receive intelligence, perhaps
from some far-distant shrine, acquainting them with the whereabouts of the
truants, and the design with which they have absented themselves. Many,
very many, die ere they can return; being exposed to many dangers and to
hardships which their immature constitutions are unable to contend with,
and are never more heard of" (447). And now I wanted more. Is the runaway, I sought to ask, a legible category to current North Indian publics? If
so, how long has it been so, and how widely accessible has it been at various times?

Slumdog notwithstanding, the multiple images of what is perceived as
"the runaway" imprints itself on varied overlapping and disparate publics,
and runaways mobilize those images in forging their itineraries and intentions. I yearned to learn what the genealogy of such a thing might be. After
all, the runaways I spoke to knew *Slumdog Millionaire* as well as I did.
Surely soon they will know *Lion* as well, or perhaps they already do. Even
though neither Jamal nor Saroo is a runaway, to runaways they will recall
or evoke a familiar — the most familiar — world. They make an image of the
displaced child alone in the city available to runaways, as a reflection, or to
would-be runaways, as an idea.

I started, there at the British Library, with the reformatory reports
themselves, which dated from three decades later than Hobbes's text and
continued to the first decade of the new (twentieth) century. In the *Report
of the Alipore Reformatory School for the Year 1880,* for example, by A. D.
Larymore, I saw many examples of released children listed as "untrace-
ables," described as "for the most part parentless and homeless," but the
notion of "running away" is largely absent. Nonetheless, some life histories
intimated or suggested the type I was looking for:

> Little Elahee Buksh, 9 years old, an accomplished pick-pocket,
> was admitted with a sentence of three years, so that at twelve
> years he will be once more free to follow his own inclinations

and I fear they will not be of the right kind. Moreover, he being under sixteen years of age, there may be a chance of his being sent here again, after a year or so, most probably with another sentence of two or three years. If such should be the case, his second state would be decidedly worse than the first, and the constraint which under other circumstances might have proved salutary loses its proper effect. His detention at the school becomes nothing else but imprisonment with forced labour, and he will leave again the school no better for having spent so long a time in it.

But let us suppose that Elahee Buskhsh [sic] had been sentenced to be detained in the Reformatory School till sixteen years of age, the result would have been one of the happiest kind. As he gradually grew up, he would have become more and more capable to appreciate the moral and industrial training of the school, and growing up towards manhood in a constant state of order, discipline, and industry, he would have been given back to the world with a predilection for that which is good and honest, and which has become, so to say, his second nature. (3–4)

Then as now, of course, as I will address for NGO approaches to "street children," socialization to a good and moralized capitalism (or at least participation in larger political orders such as empire) is framed as the solution to such "problem subjects": "Every youthful offender sent to a Reformatory School should be of such an age as to be at once practically brought under the influence of the moral and industrial training the school offers" (4). I will explore the nature of such reformatories and their relation to the present in much greater detail as well.

Only once in the records possessed by the British Library on juvenile reformatories in India is running away referred to, but it is then referred to abundantly. This appears in 1888, exclusively in the records for Alipore and Hazaribagh. In a table on "Conduct of Released Boys," under the column of "Magistrates [sic] Report on Character," the boys Khantara, Panch

Cowri, Bhagwan, Shew Narian, Dhara Nutt, Sindhoo Sing, Atibal Dome, Jan Mahomed, and Bhagwan Hazam are all given the classification "Left his home," that is, *after* release from the reformatory (though they were probably in the reformatory directly or indirectly because they left their homes before). A boy named Behari is listed as "gone to Nepal."

Yet the public audience for such things was limited, even before they became archives in the formal sense, and that limited audience was not at all the one I was interested in, so I continued my search, now trying to hone in on more widely accessible narratives, narratives around which the paradigms that motivate action are formed. What kinds of things, I wanted to know, might form paradigmatic stories of leaving home, stories that either reflect or produce models for action? I am not intimating that such things *cause* children to run away but rather that they might reflect narratives which at least attest to the presence and availability of cultural models, hint at them.

I stumbled at length upon a little-known novella from 1946 dealing with the story of a child runaway, Shibram Chakraborty's *Bari theke pheliye* (The Runaway). Chakraborty, who died in poverty in 1980, had himself run away from his childhood home in Malda District of West Bengal, which remains one of the most active points of child departure. The famed Bengali director Ritwik Ghatak in time came across *Bari theke pheliye* and, seeing in it some of his own story, adapted it into film for a 1959 release. Ghatak too had run away in childhood and had in writings dealt often with runaway characters. In *Bari theke pheliye,* village boy Kanchan dreams of a big city that ultimately disappoints and alienates him. The film ranges widely across the social landscape of 1950s Calcutta, taking in

> railway tracks and the local trains that daily bring to the city
> an incredible number of commuters and homeless displaced
> people looking for a livelihood, iconic Howrah bridge: the
> gateway to Calcutta, maidan's mounted police controlling foot-
> ball spectators at rampart, slums, ships at Princep's ghat, cows
> and ambassadors keeping company, beggar mafia, hawkers,
> nostalgia-filled office adda, pickpockets, michhil (political
> rally), glowing advertisement boards at night and the dark alleys

during the day, conjurers, day laborers, biyebari (wedding re-
ception) dinner and post-dinner impromptu classical music
jalsa, the gentlest of class warfare between English-snob city
kids and the outsider village boy, and the fiercest of survival
battles where homeless street dwellers, barely surviving, chase
dogs away and scrape for biyebari leftover food in dustbins.
(D. Chakraborty 2006)

Kanchan's departure from home is not motivated by misery nor by poverty,
and in the end, he is able to go home. But in the middle, between departure and return, he embodies an archetypal runaway's experience and finds
maternal care in a slum-dwelling woman who identifies him with her own
runaway son.

Why so much sadness in this city? asks Kanchan.
This, a worker replies, is Calcutta: it cares for no one. Go
away, go back home.

But though *Bari theke pheliye*, as I suggested, *reflects* runaway histories, it is still not the kind of cultural text I was seeking, even if it attested
to the runaway as a widely accessible and paradigmatic figure. So I considered yet other films: *Salaam Bombay* and *Slumdog Millionaire* prime
among them. *Slumdog* is, after all, a foundational cultural text, and widely
available. Many children I worked with had seen it. In the first of the pair,
Mira Nair's *Salaam Bombay,* Krishna, the typical "street kid," is depicted,
quite accurately, as a runaway (though the plot itself ventures for a time
into a jealousy-driven knife-murder bonanza).[1] Krishna's story is captured
beautifully in his narration of a letter to his mother, in a village whose actual
location he doesn't know, to a scribe, who tears up the letter after Krishna
leaves. *Salaam* is admirable in its refusal to soften or mitigate the runaway's
misery: we are left, before the credits, with him weeping silently on a back-
street stoop, tears drawing tracks in his dust-caked cheeks. But in the end,
Salaam is an art film and not highly visible to film-going, video-watching
publics, especially to children such as populate these pages.
Slumdog was something altogether different. I am asked about it all

the time, and indeed in the United States for a time it was the most accessible trope by which people could find some familiarity with what I am talking about. Is what you do, they asked, like *Slumdog?* Danny Boyle's blockbuster certainly represents an amalgam of a number of real experiences, capturing garbage scavenging, kidnapping, forced begging, orphanhood, train-riding urchinhood, and urban aspirations as it does. But it is, of course, also hyperbolic, impossible, and superheroic, and it is indeed an amalgam even a portion of which never converges on a single life. The most captivating image, however, as I discuss soon, is the one in which Jamal, the hero, and his brother Saleem, ultimately the antihero, bind together a whole nation by living on moving trains. More anon. And we shall see if *Lion* itself will become, like *Slumdog,* an available text on which to think, talk, and act, whether it will penetrate that deeply into its own context of representation.

Lion, nonetheless, which was also widely viewed in India, is about a lost child, not a runaway, and *Slumdog* is about orphans. In their engagement with their protagonists' early separation from home, both these films depict child peril at the hands of multiple types of lecherous traffickers or other abductors, a peril in which police complicity and public indifference are central. Nonetheless, it is worth considering the variable interplay of different types of childhood dislocation as they intersect with audiences and their collective enthusiasms and expectations. Richie Mehta's *Siddharth,* for its part, features a child simply trafficked away from an informal three-week employment arranged and approved by his parents, and their resulting desperate and sad search. In the wake of these films, along with the Nobel Prize awarded to Kailash Satyarthi and Bachpan Bachao Andolan, and the three Nobel nominations for Concerned for Working Children in Karnataka, much popular attention has (rightly) been paid to disappeared children and to the phenomenon of trafficking.

In terms of accessibility to publics, and related modes of concern and engagement, running away is, for its part, more morally ambiguous to the world at large—there's less of a readily visible bad guy (and who or what that bad guy is would be hard for the public to identify), and there are fewer potential heroic responses, and going home is not as easily legible as the straightforwardly unequivocally "good" solution; running away is for these reasons not as available as a core narrative of or cultural trope for tragic or happy-ending tales, even though it might more often end happily than in the case of trafficking. Nonetheless, and despite the ease with which a boundary might be drawn between running away and trafficking as two mutually exclusive modes of dislocation, the world of runaways intersects extensively with the world of child slavers. "Enslavement" might be a preferable term to "trafficking" in situations where a person is bought, sold, or owned; indeed, "trafficking" could sound deceptively benign, as a term that focuses on the movement or trade involved with the theft and acquisition of children but not the state they remain in afterward. It is worth heeding the ways that enslavement can be *masked* or *excused* by discourses of running away. A 2017 *New York Times* article entitled "Why Do So Many Indian Children Go Missing?" (similar articles also appeared in both the *Guardian* [2017] and the *Washington Post* [2012]) describes a police officer's reaction to questions about a village girl's abduction: "He [the police officer] didn't think they could have done anything more. 'Girls run away,' he said, with a shrug."

The Bengali man born in 1937 in British India, Professor Tirthankar Bose, who had first told me about *Bari theke pheliye,* also pointed my attention to a very different form of cultural model or paradigm available to village children, but one that is nonetheless also watched, viewed, and a spectacle of sorts: the many varieties of itinerant, wandering mendicant, peripatetic minstrel, ascetic, and troubadour who wander in abundance through the countryside. From Islamicate traditions there are various forms of wandering for replication or inspiration: *faqir, malang, darwesh,* and *qalandar.* From Hindu traditions, figures popularly described as *sadhu, yogi, sannyasi, naga baba, rishi, tantrik, swamiji,* and so on (but most frequently *sadhuji*). Some of these emerged historically out of the Śramaṇa milieu. From the peripatetic classes, furthermore, there are occupational and ethnic groups such as Jogi, Nath, Banjara, Gaduliya Lohar, Kalbeliya, Kanjar, Changar, and Qalandar (the ethnic group, not the mendicant type), many of them among the former "Criminal Tribes" (as defined by British law in 1871) now called "Denotified Tribes" (see Radhakrishna 2001) alongside some (sometimes) truly nomadic and seminomadic pastoralists like the Gujjar or Bakkarwal, the Gaddi, and the Rabari. In certain cases, such as that of some Jogis and Faqirs, where the professional tradition of an

Jogi/Nath mendicant child, Chhatarpur, South Delhi

ethnic group involves self-presentation as religious mendicants, the peripatetic nomadic complex intersects with the ascetic or mendicant complex in a single identity.

The relevant fact is that these people *pass through:* they beg, they perform, they stop and camp out and move on, and even among the ascetics, they often include children. Amir reflected on this:

> JONAH: Do you remember anybody, like, anybody from outside coming to your village . . . were there faqirs or sadhus who would come through?

AMIR: Yeah, yeah faqir would come, faqirs would definitely come just to beg—what do you call, I don't know what you call begging—like the way they survive.

JONAH: Yeah, like alms, almsgiving. *Zakat.*

AMIR: Ah yes zakat. Also they would just do some kind of prayer for people, like for their well-being, and also, like you know, my family was very superstitious, so if anyone would come to the door we would at least give them something, never let them go empty-handed.

JONAH: Were there kids among the faqirs?

AMIR: Yeah there were some kids, like mostly I remember like older people but sometimes they would come with their kids too.

But these formally itinerant children are sometimes already separated from families to which the group or troupe has no kin ties, which reflects the possibility that these groups *may be joined,* that kids among them might just be wandering—not fixed—individuals, nomadic, out-of-the-mold. For sedentary village children, the presence of children among such visible itinerants is an interesting, captivating, and perhaps inspiring feature: could that be me? "We grew up," recalled Bose, "with tales of children running away from abusive relatives, running away with minstrel groups, running away with mendicants, being lured away by tantriks, and very rarely and from the 1950s on running away to Bombay film studios." Tantriks—wandering devotees of the ecstatic practice of tantra—emerge out of this milieu. Sometimes seen as "occultist" practitioners of highly unorthodox sex rituals, they are infamously storied for abductions and tales of child sacrifice but also known as healers and oracles and ones who might cure ailments psychic or physical, free a sufferer from nightmares, tie amulets, and so on. Note that the narrative here highlights "stories."

Every year hundreds of child sadhus are initiated through an induction ceremony at the Kumbh Mela festival, primarily through a particular *akhara,* or order, where they are given specialized sadhu identities in the presence of other recently inducted child sadhus. These initiate children, unlike adult renunciates, may opt to wear *langot* loin coverings in place of assuming complete nakedness; once initiated they become *tangtodas,* and they may keep their naga baba status for life or for shorter preordained periods. This formalized process notwithstanding, however, a child may also join wandering sadhus (and probably Sufi mendicants, darwesh, qalandar, malang, and so on, or, in certain regions, Buddhist monastic orders) in a less institutionalized, locally specific process, more or less on the spot, with or without family involvement in the negotiation. The renunciation or continued involvement with kinship, a wider core theme in this book, is a fundamental feature of the child sadhu process: adult renunciates can perform family members' death rites in the *pind daan,* whereas child renunciates must forsake kin ties utterly and cannot perform the pind daan.

In fact the interaction of the Sadhu complex with childhood is long and old, and not always benign. William Pinch has demonstrated the ways that the "mystics who come to the Kumbh [the Kumbh Mela pilgrimage gathering] are part of religious orders that were once mercenary armies that terrified parts of northern India centuries ago." The "ghosts of armies past" who were "often employed as assassins," they "often stole or bought children to fill their ranks" (quoted in Harris 2013). In *Warrior Ascetics and Indian Empires,* Pinch (2006) illuminates the process by which these children could be torn from their families and rendered available for sale: calamity and crisis, most often in the form of famine or flood. It is by virtue of just such mechanisms that history can be translated into intimate geometries of feeling, into the course of a life, the life of a home. I will deal at length with what the archive and ethnography suggest about the role of calamity in generating regional patterns in running away, but later.

This is not an effort to create a direct and causal connection between running away and Indic mendicancy complexes; to claim that the presence of ascetics wandering through villages might plant the seed for the notion of departure is one thing, but a suggestion that a culture of running away is a distinctly regional outgrowth of mendicancy is impossible to substantiate. The most that might be said is that models for a mobile life have long been available in rural settings. But that is still saying something.

To try to dig for the cultural models that might underlie running away, and to probe the sense of sharedness and solidarity that might reside in their common experience, having heard in the field many enigmatic words for "street child," I looked to a peculiar sort of circulating text: the children's own words for their own type, and in particular the terms they use to describe themselves and their peers. How much, I wondered, could be extrapolated from these terms? Again, only tenuous proposals can be made with such material, but the implications are intriguing, and they hint at a profound complexity underlying the figure of the runaway, the urchin, or the orphan.

In North India, three primary terms for "street child" are in daily circulation: *aawaara, lawaaris,* and *kangaal* (sometimes in compound conjunction with *baccha,* "child," for example, *lawaaris baccha*). The historical etymologies of these terms are, given their referents and the contexts of their usage, captivating. In general, these terms, despite the more specific denotations suggested by their morphologies, are abstracted from situational type, so any of them might potentially be used for an orphan, a runaway, a street-working child with family nearby, or an abandoned child.

Of the three terms, only kangaal, for which Platts (1884) gives "poor," "wretched," "miserable," "indigent," "beggarly," and so on, appears to be of unambiguous Indic origin. Platts provides a historical derivation of Prakrit *kankalo,* "skull," and Sanskrit *kangkala,* "bone garland" or "skeleton"—implicitly denoting a form that is already dead or half dead.[2] Given their heightened susceptibility to death, much might be made of the application to street children of a term that ties them, as animate skeletons, to living death, and of the image. Bone children. Hence my title.

One can feel the contours of Arabic in the shape of the next of the terms. The colloquial denotation of lawaaris, whose formal translation in Hindi is "unclaimed," is "orphan" or "waif." The term's compound derivation is from Arabic: *la-,* "no" + *waaris* (or *waarith*), "successor": thus, fundamentally, "kinless" (and at the end of a lineage). This again points poignantly to something fundamental in the calculus of child vagabondage: kinlessness, both in the sense of visible accompaniment by family, as image, and as reality. Kinlessness, alongside a complex relationship to family, thus figures prominently both into the empirical features of the life course of the children who eventually are assigned the "street child" identity and into their perception by wider publics. Kinlessness is a core trope of runaway self-fashioning, and indeed it has a certain dialectic with kinship that renders the runaway socially and publicly legible as a certain category of person.

The third term, aawaara, is of uncertain origin, though the derivation is likely Persian. Platts (1884) gives "separated from one's family; without house and home; wandering, roving; astray; abandoned, lost; dissolute" and "vagabond." To try to discover more about the term's origins, I surveyed some of the world's leading Indo-Iranian philologists, many of whom rose to my request with rich answers or speculations. Claus Peter Zoller, Richard Frye, and Edmund Bosworth concurred with the Persian origin. Berkeley's Martin Schwartz suggested "the word is unquestionably Persian," but of unclear etymology; Nicholas Sims-Williams said the same. Schwartz gave a meaning of "vagrant, displaced, exiled, outcast, mistreated." Eden Naby Frye pointed out that it is attested in modern Aramaic, beguilingly explaining, "I am a native modern Aramaic speaker from Northwest Iran and *aawaara* was/is a common term in my vocabulary. It is definitely negative and ranges in meaning from astonished to confused. Oraham's dictionary* p. 12 col. b. defines the term as *wandering; a wanderer; a vagrant or listless person.* I would use it thus: *A danta'd chashmu shmidlun, pishle avaara.* When his glasses broke, he became confused. Or, the children became *wanderers,* going from village to village when their parents died." Thus a linkage with itinerancy. Christopher Brunner, further, points to the word's emergence specifically in New Persian.[3] Elizabeth Tucker asks, "Is there any chance that the word is *awaara* or *awaaraa* with short initial *a- ?* . . . If so, according to Turner, these words continue Sanskrit and Pali *apaara-,* Prakrit *avaara-* 'boundless, without boundaries.'"

This is all captivating in a literary sort of way, but in other ways as well. What is notable about the results of this scholarly concordance on the long trajectory taken by the names the children use for themselves and each other is that they assign the children to timeworn, historically circumscribed categories, categories that have clearly—along with their names and accompanying tropes—been around a long time. It would be difficult or impossible to trace, but it is not likely that the use of such terms to describe such children is new. Moreover, it is remarkable that the terms name features of these children's lives that are basic, fundamental, and true elements of their stories and trajectories, and also of their social valency: vagrancy, vagabondage, mobility, mendicancy, the experience of being lost, abstraction and separation from kin, existence at the edge of society, proximity and susceptibility to death. The lexicon reveals that such children, or at least idealized views of such children, are relatively well rooted in certain cultural paradigms. It is unclear why this is the case, or how it came to be, but reading into the implications of these names proves at the very least a fruitful exercise, a way to frame the runaway's experience and speculate or meditate on its historicity.

This question of shared and circulating models is a core problematic of this book. What are the relative roles of cultural practice and political economy in promoting child departures from home? If we look at a map of the frequency of the practice of running away by district, we see clear variation by which some regions produce far more runaways than others. This does not quite seem a direct product of poverty (that is, runaways do not come overwhelmingly or predictably from the very poorest villages or districts), nor in particular in any patterned way of proximity to cities. This calls for a careful negotiation of the boundary between tradition, on the one hand, and histories of states, markets, and empires, on the other—and a recognition of the imbrication of these two domains.

In this section I have asked what models the *public* draws upon when it imagines a child runaway and writes policy and narrative from that imagining. What models do children draw upon from the collective stock of shared images and life courses when or if they choose or are impelled by circumstance to run away? It is not simply that the runaway is imagined as a paradigmatic type by various publics, but simultaneously that the potential and real runaway, the village child, both before and in the actualized journey, imagines these imaginings, knows them, molds them, engages them critically and creatively, fashioning new ones from the antecedents they provide. But where do the imaginings come from? The relevance of all this is that it sheds some light on the larger question we are chasing here and allows us to ask: could running away be the realization of a cultural model, widely circulating and easily accessible, of possible life options, or is it a structural response, or both? It suggests, at the very least, that models for child mobility and departure have long been available.

Out Selling Children: Precarity and Histories of Child Separation from Kin

Hints and traces of the sort of precarity that lends itself to the dislocation of children, to the extreme strain that could cause children to be enslaved or sold, or to choose to leave their families for an arguably worse life, can be found even in the deep past of those child sadhus I discussed earlier who still range the land, a kind of child protomendicant, for their ranks were long ago filled by children whose families intersected deep crisis, probably not unlike the families of contemporary runaways. Unlike many other sorts of precarious, lost, or sold children of earlier times, furthermore, their past is relatively well documented. "All that was needed," reports Pinch (2006), in order to develop the massive army-like bands of yogis whose ranks included—in stark contrast to the western public's image of the sadhu—*purchased* children, "was a ready population of 'recruitable' boys, a requirement that could not have been difficult to meet in an agrarian region dependent on uncertain rainfall and lacking in any widespread social safety net in times of seasonal stress. Seventeenth and eighteenth-century north India was just such a place. The 150 years between 1554 and 1704, encompassing the reigns of Akbar, Jahangir, Shah Jahan, and Aurangzeb saw four major famines and numerous more localized dislocations and general food shortages" (80). As then, now too. Pinch's analysis lingers quite long on this connection of childhood precarity (and vulnerability and susceptibility) and wider historical calamity.[4] "Clearly there was no shortage of agrarian dislocation in Mughal India. A typical observation is made by the seventeenth-century Jesuit historian Pierre du Jarric, describing the visit of the Mughal encampment to Kashmir in 1596–1597: 'Whilst they were in the Kingdom of Caximir there was so grievous a famine that many mothers were rendered destitute, and having no means of nourishing their children, exposed them for sale in the public places of the city'" (2006: 80–81). Pinch emphasizes that the selling of children to some form of assemblage or land-owner was not at all uncommon in periods of calamity, especially in the 1700s; the human realization of this process was a substantial population of "domestic and agrestic" slaves (81). As for slavery in the late medieval period, Levi (2002) suggests the presence of children exported from their homelands and *out* of India to distant Muslim Central Asian dominions in "vast numbers," sometimes trafficked by slave-running middlemen who bartered them for pack animals in the arid far northwest: "The slave popu-lation," he writes, "was significantly augmented by the considerable num-ber of children who were sold into slavery by financially destitute parents, a factor that increased dramatically in times of famine or other economic hardship" (284).

Indeed, such things persist long, even in famines of barely more than a century ago, for example, when it was noted that "on any day and every day, mothers might be seen in the streets . . . offering children for sale" (Digby 1900, as quoted in Davis 2001). Davis, in *Late Victorian Holo-causts,* notes multiple such witnessed miseries: an account, for example, from 1879 (Osborne's) that highlights mothers selling "their children for a single scanty meal" (Davis 2001: 53); an 1897 observation in the *Mission-ary Review of the World* that "children, an intolerable burden, are sold from ten to thirty cents a piece" (152); and a missionary reporting (in 1911, but probably in reference to 1900) that parents, savvy to those outsiders' zeal for saving souls, "repeatedly . . . have offered me their children for sale at a rupee each, or about thirty cents. And they love them as we love our chil-dren. Children are now being offered for sale as low as four cents each, for a measure of grain" (171). It is not, in such instances, exclusively that an a priori "rural collapse" is to blame, but rather that such manipulations them-selves in general *create* the rural collapse by more actively tearing apart so-cial structures in the pursuit of various capitalist and precapitalist forms of accumulation, labor, and symbolic power. Child slavery, for example, was thus not just the result of the structures that enabled it, of a *system* licensing it, but also of desperation and choicelessness among the families that were motivated to sell children, and the exploitation thereof that constituted the relationship between those two realities.

Such systems persist, of course, and in force. On the price of children now: Siddharth Kara (2012) cites two traffickers in Bihar who in the cities get $90–$110 for boys, usually for carpet labor, and $155–$178 for girls, usually for sex work; the parents get a cut, of twenty dollars or 20 percent, depending on who is reporting (and probably when, and where), plus a monthly (monetary) portion of the fruit of their children's labor. These are certainly slaves. Kara does not identify a clear calamity here—what the calamity is, the vast scope and scale of global economic precarity, should emerge quite clearly in these pages—but the presence in such acts of desperation is undeniable now as it was in the past, as is the notion that they themselves become calamity.

To Davis, as to Sen, such precarity and death by calamity (and these consequent expressions of despair) are neither incidental nor determined primarily by natural process but rather they come at a moment when the stricken lands' "labor and products were being dynamically conscripted into a London-centered world economy. Millions died . . . in the very process of being forcibly incorporated into its economic structures . . . in the golden age of Liberal Capitalism" (Davis 2001: 8–9). Speaking of the same period, Marx (1867) notes, in *Capital* (vol. 1), the prevalence of "child-stealing and child-slavery for the transformation of manufacturing exploitation into factory exploitation, and the establishment of the 'true relation' between capital and labour-power" (VIII.XXXI.18). Here, for my part, I want to assert (as has Paul Farmer in the suggestion that AIDS victims in Haiti can effectively quite literally be the victims of a murder whose form took shape in sugar slavery three centuries old) that the emotional crises of today's rural Indian households, the intimate predicaments that cause such misery that eleven-year-olds will leave their mothers and fathers and beloved little siblings behind, had their seeds laid in just such moments (and in disastrous events to follow), from Green Revolution land redistributions (profiting exponentially those already able to buy pesticides and achieve high yields) to debt slavery. Those places and peoples that were most fiercely ravaged by the subjugations of empire, like much of Brazil's or Haiti's Afrodiasporic populations, are subjects of structural disadvantage so severe that even two hundred years is not enough. For the precarious, there is no backup system, no safety net: in the absolute absence of resource and recourse, a single pathogen suffices to kill; when structures are already fragile, all it takes is a touch to make the tower collapse. I will have much more to say about such constellations and realities of labor, and about dynamics and meanings of death, later, but for now, in this discussion, it is worth considering what Amir has to say about precariousness and exposure to death not only by pathogens but by exposure to something else, an all-important poison of these children's environment:

JONAH: So, uh, I wanted to ask about other, one of, in one of the chapters of my book, obviously you're not like a typical example of this but in one of my chapters I want to deal with, um, the long-term trajectory of kids who have run away, obviously, again, you are not the typical long-term trajectory, but, like most of the kids, what happens to other kids you knew in the longer term, where are they? Are they home, are they dead, are they, like what's the long-term trajectory of these kids?

AMIR: If I do a projection on their lives, the probability of surviving gets really less. Some, you know, it depends, like the five fingers are not equal, so everyone gets into their different categories, like some get into drugs, they die, some get into different kinds of activities, they die, or some may become homeless, they lose their sense, they get some brain disease, so those kinds of things happen, very few people come out from like those situations and lead a normal life.

JONAH: Right, right. And, um, what about specifics, any specific stories of kids that you knew?

AMIR: I mean, did you meet my friend D——?

JONAH: I don't think so.

AMIR: So like me and him were both railway kids at the same time so his life story is totally different than mine because, you know, he got into drugs and all, another friend, he died. He just

died recently and that I think like . . . four or five times, finally
the organization got tired of it because, you know, he doesn't
want to give up our lifestyle, he doesn't want to follow the processes, and he gets like more involved in drugs, and I mean, life
was just kind of sucking for him, anyone could die and anything
could happen, but, that's like, that's the way it is.

JONAH: Right, like how about, these kids who die, I mean,
do you think, so for example, I always heard tuberculosis rates
were very high but, I mean, what do they die of? Do they die of
HIV? Do they die of drug overdose?

AMIR: I think, uh, it could be any type of, like, it could be
anything, I mean it could be HIV, it could be just overdose of
drugs, it could be just like, you know, not a particular way of
dying, they could die from anything.

JONAH: You think they're more susceptible to dying than
other kids? Obviously, yeah.

AMIR: Yeah, yeah. The highest problem for them would be
they die, the highest chances of dying for them is substance,
drugs and substance. I mean, like, sexual activity is a different story because that takes some time to, like, build that kind
of disease in you, but like *drugs* is just, like you know, certain
drugs they kill you right away.

These children have nothing standing between them and such deadly
drugs. Once they are handed the rag, there is little hope. "I did everything that was out there," said one returnee, reflecting on his time on the
streets. "My friend used to take me to this snack shop where the shopkeeper secretly gave *solution* to us. He said it is good; I tried it and liked it."
If you are precarious to begin with and your world anomic, you've got nothing left to lean on, and a sole catalyst—an illness, a solvent to which you are
addicted because you are precarious already, in the global political economy of mental health—can tip the scales; as Marx observes, the laborer has
only the potential product of the body to sell. These children have not even

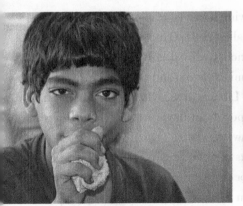

Rag children: Akash (*left*), from Gujarat, is the first runaway child I met from that more affluent region and the first to claim he had run away from an abusive mother. Mustafa (*right*) had been living in the slum at the Yamuna Pushta when it was cleared. He had run away from the resettlement village to which his family had been moved.

that left from the ascendant choices in Maslow's proverbial pyramid; they have no labor to sell, so they sell their actual bodies (as opposed to their bodies' labor product) or other people's garbage (and they are themselves seen as little more), or they seek handouts motivated by systems separate from accumulation (like Sufi or Sikh tenets of caring for the poor). The debt servitude levied heavily then and now on the people of these areas is certainly a primary motivator of childhood precarity. Today's runaways are the heirs of yesterday's privations, of what Davis (2001) calls, with some qualification, "holocausts." Davis notes that such calamity "began to unravel the tightly woven fabrics of family and religious life" (171)—encapsulating very well precisely the kind of unstable conceptualization of "breakdown" that is under question here. And he identifies in this moment of crisis plus colonialism, further, the seeds of certain forms of missionary-imperial imperatives of "child rescue" that serve the runaways even now.

History's Ghosts

What do medieval child slaves or children sold in famine two centuries ago have to do with nonstarving late-modern child runaways? Recall that the question is whether children leaving home is or is not a product of a certain intersection of disrupted social fabrics and more intimate scales of family and self. To chew on this question, to dive deeper into the triangular constellation of calamity, colonial administration, and contemporary child departure, I went back to the British Library to see what those gazetteers, unlocatable anywhere else, might report from those years of the districts from which I have seen so many runaways now.

Indeed, such long-ago moments of crisis fostered the various forms of desperation that correspondingly permitted various forms of debt bondage (that is to say that the desperation rendered the landless easily preyed upon) whose living legacy in current poverty stresses some of the families to the point that their children end up wanting to leave, but all this has been compounded by a great deal more—conflict and new forms of exploitation included. It doesn't mean that such earlier conditions *caused* current ones; it could, nonetheless, mean that there are threads linking past and present. I make the assertion here that the kinds of situations that motivated parents to sell children then are not dissimilar and also not historically unrelated to those which make families so stressed *now* that children decide to flee. On the latter—historically related or not—it is compelling, if nothing else, to see them happening in the same places. If they do, then a big question arises: why those places? The soil is not particularly poor, for example, they are not more flood prone than other areas with fewer runaways, and they are adjacent to similar areas that do not feature the same problems; indeed, even if they did, the point is that we would still be contending with Sen's assertion, which draws our attention to the role of subjugation or exploitation in *causing* calamity to produce despair.

But recall again that the purpose here is to try to make sense of the differential role of cultural and historical motivators in running away and, more generally, in the production of a childhood (or domestic) precarity that, I suggest, underlies running away, again shy of causing it. One thing concentrated clusters of child dislocation (earlier through selling, say, and now through just leaving) might suggest is tightly packed zones within which the idea of a possibility to run away circulates; another thing it might suggest is particularly vicious domains of exploitive relations; a third: both together in dialectic rapport.

Back to those musty records in London: the 1909 *Imperial Gazetteer of India, Provincial Series, Bengal,* volume 1, covering many or most of the districts the runaways of this book come from (including in Bihar), notes that for every hundred people involved in cultivation in the Bengal of a hundred years ago, no less than eighty-nine were "rent-paying tenants," nine were "agricultural labourers" (that is, they were landless and had only labor to sell), and two "live on their rents" (that is, they were landowners) (57). However, it qualifies this observation with an extreme for the Patna division (the area from which most contemporary runaways — and even trafficked children — are drawn) of 16 percent agricultural laborers across the population (with a low extreme elsewhere of 2 percent), and with the claim that "the agriculturalists are far better off in the east of the province than in the west" (57). So a vague but not irrelevant past-present connection can already be discerned, even despite the rupture that also separates them. But these records reveal an interest in the peasant's progress, under British salvation, from the depredations of moneylenders and landowners. Such language was of course meant to vaunt Britain as a force of enlightenment. In Bihar, however, claims the 1909 *Imperial Gazetteer of India,* "especially [and by contrast to "Eastern Bengal," where things are better], the tenant is still very much at the mercy of the landlord" (70). A set of measures is provided; it shows particularly high rents in "North Bihar"; simultaneously, "wages of all kinds of labour are lowest in Bihar and highest in Bengal" (71). So we can discern a narrative of a loose link, here, between complex configurations of family misery in times of crisis, and *particular* and *arbitrary* exploitive social relations. Of course, this narrative could have been used to justify replacing older exploitive social relations with new ones.

The authors of the *Gazetteer* suggest that Bihar's susceptibility to the depredations of flood and famine, despite its poverty, is greater than Bengal's, given the fertile soil: in this land, they write, of "bountiful crops . . . the population is dense, *wages are low and rents high,* and when the rains fall the distress is great. This is the zone described by Sir Richard Temple as 'the blackest of black spots on the famine map.' There has scarcely ever been a year of distress or scarcity in any part of Bengal when North Bihar did not bear the brunt of it" (102). It is worth wondering at the fact that "North Bihar" is now the very epicenter of India's childhood migration.

Closely related sentiments—eventually concluding with some imputation of local culture in order to cast British reform into the best possible
light—are reflected in L. S. S. O'Malley's (1905) *Bengal District Gazetteer
for Muzaffarpur* (again one of the *districts* with the highest numbers of contemporary runaways):

> Writing in 1877 the Collector for Muzaffarpur described the
> people as pinched and stinted, partly by reason of over-popu-
> lation and partly through the *thikadari* system and the insuffi-
> cient protection the rent laws afforded the ryots [also *raiyat* or
> *ravat*, peasant or tenant farmers]. In good years the majority of
> the ryots, he said, enjoyed a bare sufficiency of the necessaries
> of life, and in years of short outturn they suffered privations and
> sunk deeper and deeper into debt.[5] Nine years later the Collec-
> tor painted their condition in even blacker colours. "Extreme
> poverty," he wrote, "is undoubtedly the lot of the great majority
> of the inhabitants of the district. The prevailing poverty is ac-
> companied by a degree of dirt and sordidness in the personal
> habits of the people, and of grinding penuriousness, which I
> have not seen in other parts of India. The circumstances of the
> lower classes have approached dangerously to the limits of des-
> titution."

Note the repeated use of tropes in multiple passages of tarnish, blackness,
and filth. The narrative ends with a triumphalist account of British salvation
in the face of such misery.

Teasing that apart from some sort of useful observation is quite difficult indeed, given the British administration's own mastery of the use
of debt servitude (for example, in Indo-African indenture [see Prakash
1990]). Nonetheless, it indicates a social landscape defined or indeed suffused by debt. And indeed, despite such claims to salvation, as Siddharth
Kara (2012) points out in his study of debt bondage in North India, Britain "very directly expanded slavery throughout India, in particular bonded
labor" (22), which was differentiated quite stringently from older or other

forms of conventional "slavery" eradicated by the Slavery Abolition Act of 1833 (and the exchange therein, banned under the Slave Trade Abolition Act in 1807). However, as Kara highlights, East India Company lands—including most of the runaway-dense territories in Bengal and Bihar—were allowed to continue to practice slavery from 1833 until the establishment of the Raj in 1858, after which debt bondage became the prime vehicle for coercive, nearly cost-free labor in India (Kara 2012: 22).

The district gazetteers also attest to already well-articulated configurations of labor, capital, migration, and landlessness, and even to the notion or presence of a basic concept of precarity—perhaps indeed because they were in the process of exploiting such relations in the creation of a new type of exploitive relation in the form of cheap labor (which they could thus claim was morally better than its rural feudal antecedent). In *Muzaffarpur*, for example, O'Malley writes, "The one class that justifies the account given above of the destitution of the people is the landless labourer. Spending what he earns from day to day, he has little to pawn or sell in times of distress. He gets no credit from the *mahajan;* and he is the first to succumb if the crops fail and he cannot get labour. . . . The first real indication of distress throws him on the hands of Government" (1905: 87).[6] A very similar assertion (O'Malley 1905: 87, again) is made in the gazetteer for Darbhanga District. We can imagine this, since it is crafted as a defense of the government's resources, as perhaps a more candid account, as it might more fully belie the realities of imperial self-interest. "Large numbers of labourers," it is further observed here, "migrate annually at the beginning of the cold weather, in search of work on the roads, railways and fields in other districts . . . and besides this, a considerable number of the adult males are spread over other parts of India in quasi-permanent employ" (O'Malley 1905: 87–88). Such sub- or para-agrarian structures as described here, of course, were and are situated in larger relations of debt, serfdom, and production, and, though putatively "labor," are also fundamentally agrarian.

The most critical subject underscored here is debt servitude, which constitutes even now a basis for slavery in India; the 1909 *Imperial Gazetteer of India* for Bengal claims that in Bihar and Chota Nagpur "the peasantry are as a class impoverished. . . . In Chota Nagpur and the Santal Parganas [at the Ganges River mouth: another runaway-dense area] the Bengali moneylender once threatened to oust the improvident aborigines from their land" (64).[7] Elsewhere it is observed that "in Champaran the tenantry are badly off, and, during the decade preceding the settlement, 1.4 percent of the cultivators' holdings had been sold or mortgaged to money-lenders. The people are thriftless, and the majority are in debt to the *mahajan*" (64).

A review conducted through this research of hundreds of records of children arrived from Bihar showed that the district of Bihar with the highest runaway density was, amazingly but perhaps not surprisingly, this very same West Champaran, in the westernmost part of the state, and previously the westernmost part of the Bengal Presidency.[8] In this analysis, the overall number of runaways from West Champaran was well over twice those from the district with the second-highest runaway density (which was East Champaran); taken together, the number of runaways from both of the Champaran districts combined outstripped those from the next-highest district (Darbhanga) by around 400 percent and contained a quarter of all the runaways in the study's top-ten documented districts. When adjusted for population, in a "runaway frequency factor," West Champaran had a runaway *rate* 136 percent higher than the closest district's, numerically speaking, which was Bhabua, in the extreme southeast of the state, due south of West Champaran.

While, again, I do not suggest causality, that there is an echo between past and present, and that patterns of peonage persist over time, cannot be denied. But the question remains: why? Why one district more than another at any time in history?[9] What creates a "vortex of suffering," a concentrated zone among many others that might be, but are not so? The answer suggested by all this archival matter is quite simple: exploitative relations of debt and consequent acute stress for landless laborer and tenant farmer alike — and, most crucially, for their children (and indeed children of course experience this stress directly, and not only through parents: they are themselves, as famously depicted by Tamil novelist Perumal Murugan in *Seasons of the Palm* [2004], too-common subjects of indenture and its attendant exacting punishments). Then there may be another question: why *there* more than elsewhere for *that* phenomenon, as well? What allows for the growth of exploitive structures in particular areas? That, perhaps, takes us to the dead-end answer of historical accident that accompanies the thread of "why" questions for any area on any particular question: Why West Africa? Why Europe? Why Syria? Or, if we wish to be a bit more reductionist, to the answer of ecological coincidence: the productivity of the soil, for example, or the way settlements happen to be arranged over land.

But if we take Champaran, for example, as a case to think on by virtue of its persistent appearance, alongside a few other districts, in the materials, hints about the *why*, if indeed it exists, may come into clearer resolution. What happened or happens in Champaran that spans the colonial moment and the now, and that might, further, be tied productively to events of child departure? It would be easy, perhaps, listening to the words of these gazetteers, to overlook the centrality of a colonial commodity—indigo—in the production of local misery in Champaran, and to attribute all inequality there to a mode of debt servitude that British rulers framed themselves as actively bringing to an end, to replace with capitalist accumulation and modern modes of taxation and earning—which again can be seen as a prescient justification for *labor* as better than agrarian exploitation—and with the cultivation of commodities they favored.

It is historically notable that one of the runaway-densest districts in what is unequivocally the runaway-densest state was the site of tremendous agrarian exploitation under empire in the name of indigo, and consequently, the district from which Gandhi's seminal first Satyagraha campaign was launched—and in protest to this particular iniquity in this particular place. In other words, it was already well identifiable as a site of injustice— enough, by 1916, to draw Gandhi's and other political radicals' attention, the nascent nation's, and ultimately the world's. Indigo came into favor, of course, as a highly marketable clothing dye, the intense European demand for which by the late 1700s motivated the Raj to expand its area of cultivation. Between the late eighteenth and early nineteenth centuries, the share of British indigo obtained from India more than trebled, accounting at the end of this period of transformation for nearly all of the indigo processed in the metropole. This led to a kind of "indigo rush," wherein opportunistic British planters saw a chance to seize land, labor, and wealth. The less

common *nij* model of plantation building required the eviction of peasants from lands around the factory, and the indigo planting co-occurred
with local rice planting. Much more common was the *ryoti* (again, also
rayati or *rawati*) system, in which tenant planters were coercively entered
into contractual relations with landlords, incentivized with up-front, low-
interest monetary reward but also required to hand over at least a quarter
of their land to indigo, and they were in fact given very little to compen-
sate for their land and labor. Repayment consisted of the fulfillment of a
single planting cycle, at which point a new "grant" was given; this trapped
them in a continuous cycle of debt that they could never fully earn their
way out of. Moreover, landholding plantation owners demanded cultivation on the highest-quality fields, which required the ceding of rice, which
in turn shifted their primary agrarian activity to a commodity crop from
a food crop, with direct impacts on food security, precarity, and survival.
Indigo's extensive rhizomes, furthermore, generated stresses on the land
that amounted to an ecological disaster, depleting and destroying topsoil
and precluding the subsequent cultivation of rice, their major staple (see
Praharaj 2012 and Pouchepadass 1999).

    As mentioned in the introduction, one of the book's characters, Nabil
(Toonda), comes from a district in Bangladesh that was in 1873 its own
site of analogous peasant exploitation and subsequent rebellion—there in
the name of jute, rather than indigo, and the heavy taxes levied by landlords over the cultivating raiyats. Tarique (2008) claims that the peasants'
struggle was "not against the 'system' and certainly not against the British
rule. In fact their slogan was 'to be the *ryots* of her majesty the queen and
of her only.'" Indeed, on the surface, it appeared that the British shut down
the rebellion by enforcing better standards for the peasants against the landlords through new legal structures. But it is easy to see a specious logic in
Tarique's assertion: the landlords were, after all, insisting upon the most
profitable mode of cultivating a commodity absolutely central to the economy of empire, and thus both directly and indirectly acting in its service.
Continuing jute cultivation without the "nuisance of rebellion" certainly
served both the Raj and the zamindars. Pabna was also active in the indigo
rebellion of 1859.

As with jute, whether or not the plantation owners were British (which they largely were), indigo cultivation served the British economy. Nonetheless, the zamindar landowners were not isomorphic with the planters, and in some cases the former joined forces with the peasants. In response to mounting peasant resistance, the planters and their deputies began to deploy *lathiyals* (or *lathials*, "stick-wielders"), among others, to wield sufficient force to coerce the peasants into submission so they would carry on with the indigo. When in 1859 Lieutenant Governor Ashley Eden visited the region and pronounced the end of the ryot's compulsion to enter into indigo contracts, the peasantry saw it as a Victorian edict for their liberation, though it was in actuality an effort to quell rebellion (which the visit failed to do). Shortly thereafter, the Raj mobilized militias to safeguard the planters and then created the Indigo Commission, essentially an inquest that imputed the peasants with destabilizing the indigo economy and reinstated their contracts, but also allowed them to opt out in the future; the rebellion of 1859 was thus brought to an end, and indigo cultivation began to falter in Bengal (Pouchepadass 1999). It was now shifted to Bihar, where Gandhi heard (from Raj Kumar Shukla, a planter) of the persistent misery of indigo planting of over sixty years. Indeed, I suggest, that misery has persisted to the current era—a misery generated specifically of acutely exploitive labor conditions. A corollary—or perhaps primary—question is whether it is more the legacy of misery or the legacy of resistance that should be tied to running away.

It was surely such conditions of debt, failing agrarian systems, and unrest that lent themselves to the expansive engagement of what are now the runaway-densest districts with global indentured labor migration under British direction. It could be argued that the preexisting conditions of cyclic debt servitude played directly into the new forms of labor debt the Raj instated for sugar, railway, and other labor in far-flung colonies, Trinidad to Mauritius, Fiji to Guyana, in what Khal Torabully and Marina Carter (2002) call "coolitude"—a process that created its own new forms of peonage and debt slavery and, further, set into place new patterns of departure. The emergence of mass global labor indenture in Bihar and Bengal is roughly contemporaneous with the genesis of resistance to indigo cultivation. Thus we have a nidus, a tangle, of co-occurring complex forces and events emerging from empire and its local incarnations, and from new iterations of capital, that seems to coincide with contemporary autonomous child migration.

In the data on runaways, especially when adjusted for proximity to any given city, no state is more visible than Bihar, which is followed by adjacent regions of West Bengal, southern and Gangetic Uttar Pradesh, and then Jharkhand, Chhattisgarh, and Orissa. The specific districts that emerge most clearly as critical sites of child departure are, in Bihar, the two Champarans, Saran, Sitamarhi, Samastipur, Gaya, Darbhanga, Patna, Motihari, Madhubani, and Muzaffarpur; and in Bengal, New Jalpaiguri, the 24 Parganas, Malda, Asansol, and perhaps Cooch Behar. That a circumstantial, anecdotal connection can be drawn between these troubled pasts and the present state of children is indisputable. We can see at the very least that areas whose social relations and fabrics are somehow historically disrupted by exploitative relations are the same areas from which we see children leaving; when asked, those children attribute their departures to family stresses emerging from conditions that we can then interpret as clearly linked to these histories. This is not causal, but correlative. To this we must add, of course, the primacy of the railway, another colonial creation: the areas with the most runaways are the areas where trains furnish the opportunity for departure, particularly on the main Calcutta-to-Delhi line.[10] As soon as one enters Bihar on this line the presence of children by the tracks and in the trains intensifies; their lives are interwoven with the railways.

And yet, in the villages themselves that were part of this research, the reasons families and children provided for running away never really attained that historical depth—they merely hinted at its existence, for instance through tying child departure into larger trends of labor migration by adults to cities, by referring to "poverty," marginal ethnic identity, or family crisis, none of which illuminates the breadth of the phenomenon, its sharedness.

Thus in Siwayapatti, Sitamarhi District, northern Bihar (in the vicinity of the Nepal boundary), the "repatriation" of the child Roshan (whom I discuss elsewhere) by an NGO became, upon his arrival, the center of a vicious, fraught, tear-soaked yelling match throughout the village, accusations hurtling from person to person over the nature of the boy's departure.[11] Roshan was apparently a serial runaway, and an upper-caste Brahmin, but simultaneously poor: his landless family served as tenant cultivators on others' land. He claimed he learned how to run away from another boy who had already done it. To the south, in remote Phulwariya, north central Jharkhand, the return of the children Ajay and Vijay yielded a glimpse into historical identities in the form of the ramshackle *phultoli basti* slum populated entirely by the little-studied ethnic Mahalis, one of the "scheduled tribes."[12] In a room shared by cattle and people alike, Ajay's mother lamented their family's grievous poverty, particularly after her husband's death. There were three runaway cases from this slum alone. In these and yet other villages in this landscape suffused with runaway stories, where every village seems to have multiple runaways and it is scarcely out of the ordinary, it is intimated repeatedly that violence between Maoist Naxalites and the government explains higher rates of running away, but we could not uncover precisely how, except to say that conditions make such-and-such hamlet less desirable to stay in. But Naxalite violence *itself* must absolutely be taken to be an offshoot of and response to these very conditions of social distress and subjugation. Indeed, crisis begets crisis, and what is at one moment a response to crisis (Naxalite rebellions, say, or an urban handgun culture) may later become its very fabric; thus indeed if running away is a reaction to anomie, then eventually it may become the very shape of the sense of anomie, one of its foundations. If earlier a tyrannical world of indigo cultivation or caste violence or new landlessness was at the root of things seeming out of balance, and children consequently run away, then in time that world of vanished children can seem the *source* of imbalance.

A mother dies, a father leaves, elders feud, and a family falls apart. The home environment becomes intolerable, the former rhythms of an everyday life are lost and recede into memory, staying on takes on a veneer of despair and begins to look untenable. The floods come, drought spreads across the land, stressing things ever further. Now the railway presents itself in all its liberatory, dreamy potentiality, the children who can be seen occupying it compel and captivate the housebound child, and a possibility of departure begins to take shape. This shape comes into clearer resolution through the stories of other children who've left and still others who've returned, having seen the grand or gritty other side, the sometimes-enticing site of desire and potentiality and adventure, or at least of the absence of the burdens of home. One night, perhaps one night when the child knows the reason for his departure will be obvious to his parents (a recent blow, a scalding feud, a public shaming), the child packs (or does not), grabs his or others' money if available, and in the secrecy of dark or the unremarkable noon hops on a train to the next depot, the small town, and then the state capital, from which point many children are gathered in a kind of entrepôt of children similarly cast loose of their own accord, most of them in transit to a larger city, and the child takes off again: Mumbai, Calcutta, Chennai, Delhi.

And when the next day dawns, what then? Something is wrong, something is amiss—that much is immediately evident. An absence, a rupture, is perceived. The sheets are rumpled, the child is gone. The child is not where he was or tends to be, nor does he appear when expected. Routine, the earth's revolution, the sun's rhythm, fails to return him to his usual place. A mother or father or sibling asks around: no one has seen him, no. He wasn't at the cricket pitch, he wasn't at school, he didn't come to the candy store, he didn't show up at Manu's house. But yes, he *was* seen yesterday; no, nothing seemed out of the ordinary. And no: he did not say a thing (or, perhaps worse, he did). His regular haunts are checked, reviewed: nothing. Quotidian anomalies—sneaking out, a day smoking *charas* in the woods, a trip to Uncle's house, a quick tryst—all these are ruled out. The previous day's miseries and fights are reviewed. Why this? Perhaps the parents know right away, perhaps they do not. Perhaps it was someone else, but perhaps not. Perhaps it was just them. Blame and accusation start to infect everyone: fingers are pointed, imputations hurled. Some people say this kid's an aberrant, not right in the head. Some think of his secrets, but say nothing. A wider family search web is deployed. The police are notified. By the next morning panic starts to set in, but surely, one decides, one is sure of it, he'll be back the next day, apologizing, saying he went to Sita-marhi or Jagdalpur with that good-for-nothing friend, with that crush he thought nobody would let him be close to, maybe to get drunk, steal cell phones, buy clothes. But when he is not back the next day, it is not only panic but terror that sets in. The child has been stolen, or has run away, or is dead. Hence, in our interviews, we heard a mother (Ajay's) say, *I have no idea why he would keep us in the dark. We were going crazy with worry,* and a different child's sister say, *Because of Satish the whole household is worried.* There are tears and true despair, and for those able to keep it together, flash-lights, dogs, and, perhaps, phone calls, eventually posters with identifying information and a phone number (sometimes posters, calls, and other such things prove impossible; Roshan, for example, said he did not know his family's phone number). Meanwhile, resignation, the thought of the worst, settles in, lingers like a specter (or was that the child's?). The wound keens. A hopeless half-mourning: even if he returns, he will be tainted, older, vio-lated, worldly, addicted, maimed—not my baby, not anymore. They have heard the stories of those who've gone before (we heard returnees who'd emulated a sister or an uncle, among others). Nonetheless, until the last, a glimmer of hope remains. And so the search goes on: depots, metropolises, shelters, and registries, and then fatigue, the exigencies of the other chil-dren to which no attention has been paid, the shame one bears in the vil-lage for being a bad parent, for failing to keep him in home's orbit, and the grinding, bleak, hopeless reality of the life and labor to which one must, in-exorably and inevitably, return. The half-ghost, that photonegative version of the child, sticks around, as the demons of worry gnaw and chew. The thing is, one feels, looking to the path that leads to the station, you never know, and maybe, one day . . . And so it is, as I will illustrate later with stories, that after weeks or years or even decades, sometimes he walks back down that path and into the door. More often, he does not.

Santosh's trajectory, as his mother recounted it, expresses an arc that is not atypical, that embodies some of these elements:

> He was about 10 years old.
> I had just given birth to my daughter and Sunil went to

out to the bathroom [one assumes here it was out of the house]. I don't know what crossed his mind; he just ran away. We spent so much money looking for him but could not find him anywhere.

He just came back on his own, after four or five years.

My eldest daughter used to run away to a relative's house without telling anyone. We got her married. Now she has three kids and she does not run away.

People said maybe some relatives kidnapped him, but I didn't believe that. I saw him go in front of my eyes, so I thought he'd just left to play. But instead he sat on a train until he got to Delhi. After he returned I asked him if someone took him there or influenced his decision in some way. But he said no, it was his own choice.

People advised us to file a police complaint, but I did not do it. In whose name would I file a complaint? So instead I tried my best to find him. And he just showed up when he wanted to come. We would have filed a police complaint if we thought we had an enemy who could have kidnapped the child. And we had no enemies, so I didn't go to the police. Moreover, I didn't have a photograph of Santosh. How would they have filed a complaint without a picture? I don't have any pictures of my kids.

Distressing: one can feel the texture, in this narrative, of the mother's desperation, the surprise at the child's sudden, spontaneous, seemingly flippant and casual departure. Her search for reasons, the lack of means, the lack of access to anything that might help — save hope. But hope is scarce in such situations.

All this repeats itself across the land.[13] And what is remarkable in it is neither its unpatterned scattering nor its universality but its complex but well-mapped clustering. Why, again, some places more than others? Why these same limited stories over and over again? This is why I turned to histories of geography and geographies of exploitation. Because the peasants of some places were more subject to servitude than others. Because labor was drawn from some places more than others. Because the land was exhausted in some places more than others. An accident, one might say, or a calculus, others might retort. What, then, to say, after all this, about the postulate that itself sparked this project—that running away must be a response to crisis, a dramatic realization of anomie in the world? It is certainly no coincidence that these children come from some of the world's great "vortices of misery."

And indeed there really *is*, I believe, something going on in the villages children run away from the most that counts as "crisis" on some level, and particularly on the level of the disruption of kinship practice and structure. The confluence of all these things over time and space—slavery and child sale, indigo exploitation, rebellion, Naxal-Maoism, debt-driven coolie labor, agricultural crisis—does itself point to the *reality* of the notion of "vortices of misery," and, as I indicate, perhaps even nested or cascading vortices, wherein each response provides its own context for new misery. Even the postulate that the villages manifest a cultural tendency, a shared norm that decrees leaving home alone acceptable and available, is likely a formation that represents culture responding to historical conditions, a structure of practice and discourse that has developed in dialogue with a catastrophic scale and pace of change.

What is it that keeps the same place in a condition of continual ravage over three centuries? And what conclusion then is available about whether running away represents a response (collective or individual) to "things falling apart," to some anomic condition?[14] I suppose it depends on what can be considered *a-nomic*. What is, or was, the nomos? How far removed is it in time from the situation we deem anomic? Our presupposition tends to hold that the romantic and picturesque thing that has fallen apart must be a thing that was good, so we assume that the falling-apart is of generative, intact kin structures, traditional ways of life, cohesive and integrated communities sharing resources. For example, if the cohesion of a system is already based on a system that to the observer seems off-kilter, out of balance, and exploitive, then it is not that things have fallen apart but rather that they were based on already unequal or unjust (following Sandel 2009 and ultimately Rawls 1971) relations of power and were already somehow oppressive, which is rather a different thing and must change our account of "what happened." If my argument rests on a generalized notion of precarity, then the assumption I make underlyingly is that precarity is not the "natural" state of things and must be a "fall" from something else, a collapse, a disintegration. Nonetheless, culture and kinship *do* hold lives and places together, keep people at least believing that they are rooted, and sometimes people *do* describe themselves as living in a world fallen apart.

Death and the Urchin

MANY OF NORTH INDIA'S street-dwelling children die. It might be the case that an observer sees them daily without giving them a thought but only hears of them, perhaps namelessly and facelessly, after a death; it is also, however—and contrastingly—the case that most street children's deaths go unnoted, unremarked, and even, at times, unknown to any for a long time. How is a runaway, a child out of place, framed if rendered as absence rather than presence? What is the interaction of the lawaaris and the aawaara with death? In this chapter I underscore the exceptionally intimate relationship the "urchin" has with death, including through vulnerability and, in the public eye, a certain dispensability, rooted in the social gaps runaways inhabit—between young and old, between life and not quite life, between citizenship and its absence.

My fundamental concern here, in this chapter, is with the social life of child death on Delhi's streets. In particular I wish to make sense of the public circulation, iteration, and visibility of knowledge, narratives, and images of street children's deaths, or—to complement—the invisibility, silence, and unknowability such passings may invoke. What is the valency of the death of a noncitizen barely visible to and barely recognized by the authors of space and the guarantors of rights, and yet highly visible to their violators? What is death, socially speaking, without traceable kin, name, place of origin, or legal existence? What does such a person's death reveal about the value they are assigned by society (see Conklin and Morgan 1996; Gillespie 2001)? I propose that the death of solo children in Delhi, and their interaction with death, reveals much about the calculus of self and citizenship in postcolonial India, and that the reality of life in postcolonial India, in turn, is inscribed into street children's encounters with dying.

Petu, Frame Zero: Aperture and Passage

One hot April day, at New Delhi Railway Station, a boy known to his friends only by the nickname Petu (Tummyboy) ran alongside an accelerating long-haul train as it pulled away from the terminal, likely hoping either to ride it a short distance before hopping off or to scavenge some plastic bottles for cash. But as Petu climbed aboard the moving train, he fell or — as some said — he was pushed or shoved. He became stuck, and as he tried to extract or right himself, maybe to climb back aboard the train, he was dragged for a considerable distance and then run over by the train. His legs were both sliced off near the groin, and he died beside the tracks as the Shramjeevi Express continued on to Bihar.

As part of one of my daily research itineraries, I came to the station some fifteen minutes later, arriving by train from Hazrat Nizamuddin Long-Haul Station, where I'd been with another group I was working with. On the way in, it later struck me, I'd passed Petu's death site by rail, myself perched at an open door, right around the moment it had happened. There, at New Delhi, I joined Khushboo and two of the older runaways I'd come to know there; following the rumor that a child had just died, we four ventured out along the tracks to the site of the death, where Petu's mangled body lay. The corpse was abandoned, its bones jutting, its tissue unraveled everywhere, and it bore signs of hemorrhage and impact on the head and abdomen. Nobody was attending it. Vultures and crows were by now circling, and flies swarming, just as they do upon the living in such places. The boy's clothes had been torn off by the same forces that had ripped off his limbs. The garments' remainders radiated out from his body like blood-soaked banners upon the stone and steel.

Khushboo called the police, who had not yet been notified. Within minutes we were standing alongside several police officers and medical workers, who were mostly staring at the body and considering what to do. I had a camera. Khushboo proposed that a picture be taken. In the fog of my own cultural anxieties over death, I was deeply uneasy: would such a thing be a violation, supernatural or worldly, of life or of ethnographic ethics? Would it produce something that could only ever be a lurid objectification—or trivialization—of life and death? Would it violate some institutional edict? Khushboo and the two boys, Arshad and Vikram, were less ambivalent. The senior police officer was curt and, refusing to allow a photo, offered a terse justification: "Take a picture," he said, "and they'll go ahead and make a story of it." We took the photo nonetheless; though I never permitted the photo's public circulation, over the months that followed we watched as Petu's death generated its own social life and public reality. It became a story.

Why it is that a story should not be made of a child sliced upon a railway line into three pieces, and that such a story might indeed be threatening or dangerous (see Verdery 1999), is among the central problematics of this chapter. What becomes of a noncitizen who to the public lacks name, kin, or legitimate space and who to the law is not even permitted to exist in this space, even in death? Is a subject who is forbidden to live in the zone of urban cleansing and aesthetic regulation that he yet and nonetheless lives in also forbidden to die there (see Ghertner 2015 for an exploration of such cleansing and regulation)? Why should the death of one child cause hundreds of others inhabiting the same area to flee in fear of punishment—not for the death itself, but for their potential to draw attention to it, and for the death's potential to draw attention to them?

Petu was, like most other children in this book, a "railway child," a runaway from the very Bihar whose past I've excavated already, the same region that the train that killed him—the Shramjeevi Express to Rajgir—was bound for on that April 4.[1] A well-entrenched resident of Platforms 7–8, and under the patronage of the young "boss" named Mental, Petu made his living, like many other children, gathering empty plastic bottles from the station's trains to resell to middlemen for recycling: an entire subsistence based on others' waste, on the empty shell of a global commodity. Petu had taken his lunch that day at the "contact point" of the charitable organization Salaam Baalak Trust and was likely planning to tarry in the train car only a short while. He was perhaps eleven or twelve.

Of course, the train did not kill him, really: through the set of mechanisms that Paul Farmer (1996) labels "structural violence," much else besides, as I've shown—history, poverty, empire, and landlessness included—brought him to the place, actual and conceptual, where it was even possible that a train could end his life in such a way. A similar set of forces brought Nabil to the moment when his arm was severed. Even Guddu, Saroo's brother in *Lion,* met a cognate fate, and in his case too it was not the train alone that killed him but the labor-rooted structural circumstances that put him on those tracks to begin with. The train was the vehicle, so to speak, but not the true cause, of Petu's death. In life, in many ways, Petu was a person with no public recognizability, a noncitizen, invisible, nameless, kinless, even, in the social eye. Upon his passing, it was nearly impossible to track or trace his "real" identity. In time we learned from Salaam Baalak's records a real name: Rohit. No last name. No traceable parents. To this day we do not know whether the parents, wherever they are (or are not), know of their son's death.[2] Is it possible that it is kin, or perhaps also legal existence, that alone render a person truly socially grievable, legitimately alive, or even legitimately dead? Is grievability tied to such forms of locatability, of attributes and trappings that are seen legitimately to ossify a bureaucratically identifiable self in society's matrix?

Judith Butler (2010) suggests that some lives are, by virtue of how they are situated with respect to the social and the historical, "ungrievable." The line she draws between grievability and ungrievability—relevant even though framed in the radically different context of Iraq's wars—can be tied to other axes of valuation: "recognizability," for example, and even "loseability," each seeming to highlight a subtly different semantic designation. Most fundamentally, however, a life's ungrievability is a product of its "precariousness," its susceptibility to dying. In contrast to Agamben's (1998) "sacrificability," Butler's grievability admits of the assessment of historically contingent identities and the differential assignment of worth according to those variable terms.

Is Petu grievable in death? How precarious was his life? Is any child in the station a potential Petu? What does it take to render him recognizable and grievable? A century before Butler's observation, Hertz ([1907] 1960) reflected that by contrast to adults' deaths, "the death of . . . a child [may] go almost unnoticed; it will arouse no emotion, occasion no ritual" (76). Petu's death has been notable in its passage through a series of mediated and legal "frames" of the sort Butler describes. Has this sufficed to

sediment the status of his death as visible, knowable, and legitimate? In this
chapter I trace the deaths of some of this book's runaway children through
the various frames they transit, just as I observe the absence of such passages. I also interrogate the way another's death, particularly that of a kin
member (but also of a companion, especially when the latter demands the
living child's labor), can frame the worth of a street child's life—how, in
other words, a street child's life may even or only gain its meaning and
value by (social) contact with the "other world," by passage through frames
of death.[3]

Another Vagabond Death, Another Vagabond Less: The Public and Not-So-Public Life of Dead "Street Children"

When runaways like Petu die in the railway station, a tragically common
occurrence, public attention is drawn to the stain on properly modern and
socially clean urban space they embody, to the failure of the state to care
for them, the failure of the city to keep them out of sight, the failure of the
self-envisioned society to attain the sleek aspirational sheen of a Dubai or a
Kuala Lumpur. These children alive are a blight on prim propriety. Dead,
they threaten to consolidate all gazes upon lapse, upon social dirt (following Douglas 1966), people out of place, wrong age, wrong behavior, wrong
space. The still-living know that with a death in their ranks they will be
punished simply for remaining in this zone of limited regulation: for officials, the press, and others will turn their attention here, and the focus will
be upon removing them from visibility: why, those concerned with the control of space will ask, are these children here? Surely if they were not, we
would not have such nuisance upon us and the city would look better for
Commonwealth Games, Olympic bids, and the like. Railway children know
they are not to be seen, and that they are objects to be cleansed from space,
in their current—kinless and vagabond—form. With kin, or with institutional iteration, they are a different thing altogether.

Though the politics of Petu's death may seem to have begun with the
police officers' arrival, in fact shortly after it, for these very reasons, well
before law enforcement arrived, many or most of the station-dwelling children scattered, as they periodically do in response to new decrees of law

or event, fearful of a reprisal that they knew well would be constituted of purging, deportation, and interrogation. Such expulsion was, in fact, enacted only days later.

But there is much to recount first: we're not there (or then) yet. We are still, in fact, lingering there in the heat by the tracks, where Arshad and Vikram are standing with Khushboo and me, and nobody is doing anything about the body. Arshad offers me a suggestion: you do it. *Move the body.* I squirm, socially speaking: I am unsure if he is joking. And then he clarifies: you and me, together, and Vikram too, let's go. I see he is serious and offer some reasons why we should not. And suddenly I feel myself on the verge of tears. Train after train lumbers by, their shadows passing over Petu's corpse. Why have all the other children disappeared? Where is the Mental who is supposed to be Petu's "protector"? Where are the station staff whom I imagine properly responsible for such matters?

Mental, as far as we know, never returned to his home in the station, and Khushboo found him in the Ganges River pilgrimage town of Haridwar nearly a year later with a new group of boys under him. Many of the other boys also stayed away that day, and those who did were right to do so. Here is why: as a letter about Petu's death and legal follow-up addressed to the Delhi Council for the Protection of Child Rights by the Commissioner of Police for Delhi Railways states, "The local police of New Delhi Railway Station took proactive steps for the safety and security of Vagabond children loitering on the platforms and station areas. 182 children were rescued and were handed over to NGO like Child Line for better care and protection. The children were produced before the Child Welfare Commission for better care and protection. All efforts have been made to keep away the children from the tracks and loco sheds on our parts."[4] Much is remarkable about this; in particular, it is clear to most readers that "safety," "security," "rescued," and "better care and protection" are euphemisms that obscure the widely known tendency of police forces and municipal agents to "cleanse" such spaces for the attainment of aesthetic objectives (see Appadurai 2000; Baviskar 2003; Chakrabarty 1992; Rao 2010), such as were pursued with vigorous force upon the occasion of the Commonwealth Games of 2010 (slum clearances, homeless deportations).[5] It is also known, as I have written, that Delhi's street-dwelling children tend to seek to avoid such charities, even if they promise rescue, identifying in them (perhaps rightly) still another mode of confinement, discipline, and surveillance. Various other terms in this text, including *vagabond*, which is, notably, capitalized in the original (and which is, in fact, the translation of the Hindi term used both by and for these children: *aawaara*), *handed over*, *produced*, *loitering*, and *keep children away* belie what can be discerned as the basic attitude toward the visibility of such children: the visibility that a death is understood to threaten to highlight. And indeed, our follow-up confirmed what we expected: that this "rescue" operation was in fact a crackdown, with the expected process of eviction and expulsion to achieve the goal of invisibility. The children were simply pushed, for a time, outside of the station's (and thus, at some level, nation's) boundaries.

Initial attempts to trace Petu's real identity failed. Eventually we (Khushboo and I) discovered that he had checked in that day at the Salaam Baalak lunch and that he was known there, at the drop-in center. To recap: he did, as mentioned, have an official "station arrival" record of registration with the NGO, which suggested, again, that his name was Rohit, and that his father's name was Shri Gurde Maldar.[6] The only address listed was the state, Bihar. Nothing else, no district, no village, no birth date, and no other identifying information. As far as we know, it has thus far been impossible to connect Petu to any known place. In fact, most Delhi street children dissimulate, as is discussed elsewhere, as part of a wider strategy of evasion and self-protection, and in this a key initial ruse is usually that of the name. Thus, if we look at the record here, it emerges as rather unlikely: both *Shri* and *Maldar* are associated with high and landed status; in particular, *Maldar* refers pointedly to a wealthy person, so the name given suggests "big shot." *Gurde* means "kidney" but is likely a mistranscription of *Gurdev*. The whole name has the improbable sound of a deception produced (by Petu) for the file and its agent(s). Most station runaways do not want their families tracked down; it is the families that they are usually fleeing. Nor do they want their villages in there; they fear being "repatriated," which they may well be.

Let me briefly digress to explain that personal names as deployed by "street children" in this way become fluid vehicles of strategic self-positioning to circumvent forms of power that seek to define, fix, and track

Petu, entextualized

the children. As flight and escape, particularly from untenable domestic situations, are itself at the very center of their stories of self, NGOs, government programs of "child rescue," and service-providing religious institutions are seen simply to replicate the sort of uncomfortable captivity from which they ran, and the common wisdom is thus to avoid them. As any agent in public space threatens to temper their mobility and freedom, and potentially to make their location known to forces they wish to evade, these runaways may use dozens of names. The names are not used in unpatterned ways, but rather they are situationally contingent. A child may vary her name by location, or by occupational situation, or by the age or other category of her conversational partner.

To illustrate this, I might point to another "vagabond" boy—this one inhabiting, among the other homeless, the park at the center of Delhi's largest mosque, the Jama Masjid. At the time we met, all the other runaways knew him, and in his long career there he had been in and out of multiple charities and homes many times. It was possible to identify at least four separate major NGOs, each of which processes thousands of cases, in which he was known—and in each he had a different name, and each a different story. I knew the boy both as (let's say) Akhil and, simultaneously (let's say), Aziz. As the story goes, Akhil ran away very young. As he tells it, he was sold off—into informal bonded labor—multiple times, first to a tea seller, then a roadside snack maker, who sold him off again, and so on, a few hundred rupees each time. In some versions of Aziz-Akhil's self, he'd originated in rural Bihar or Uttar Pradesh. In others, he'd come from within Delhi's limits. He'd run away from each NGO he'd been sheltered by. A Hindu social worker at the NGO Butterflies registered him as Akhil because she did not want to give him a Muslim name, and Akhil stuck as his sometimes-name.

Similar stories emerge all over the landscape of street-dwelling children: a girl in New Delhi Railway Station with at least four names, most of them carrying a sort of mythic texture, names of gods' consorts and so on. And indeed children throughout India, and across South Asia, invariably dissimulate, at least initially. The now-adult former runaway Amir was ashamed to tell even other children why he'd run away, once on the streets:

JONAH: Did you ever tell this story once you got to Delhi or did you not tell other people that story?

AMIR: No no, this was very personal and didn't want to like tell lots of people what was really going on so I said I ran away, but you know, that's like the realistic, I mean, I didn't really feel like telling them that exact, you know I was like, I became really bad, I started stealing things, I had a bad image, no, I just said simply that I ran away from my home that's all.

JONAH: Do you think, if I go through Delhi, like today and I say why did you come, why did you run away, people will tell me some story but most of the story won't be the whole, you know . . .

AMIR: No, no, no just because they are survivors, they don't want to tell you what really happened to them. Now that I'm mature and have like gone to school I'm telling you the truth but maybe if I was not mature or not going to school maybe I would not have because those are scary parts to tell people, I don't know, that's how I feel.

JONAH: Well that's why I'm interviewing you in particular, I mean, you know, because I speak to all these kids and they're like oh, just like that, oh I just came, my mother hit me . . .

AMIR: They are not being true to themselves a hundred percent, they are like telling you truth maybe fifty percent but they are not telling you truth fifty percent.

JONAH: But maybe that's their—

AMIR: I said that's how you have to see that, like you know, if they're young kids they're not going to tell you the whole truth.

JONAH: Well you know and maybe that's their defense, maybe that's their way of protecting themselves, they don't want to be found by their family.

AMIR: That's exactly, yes, you have the right thought in your mind.

And why is it that Akhil-Aziz—and likely Petu too—did not want his true name known? What threat does that pose? He probably knew quite rightly that dissimulation and deception kept him untrackable, unlocatable, and that name and origin threatened to send him, well, home, or somewhere else where he couldn't move as he wanted. Akhil-Aziz did not wish to occupy registers and logs, did not care to be put in a home, did not want to be surveilled across the land, did not want to inhabit a line in Matthew Hull's (2012) proverbial File. Wherever it is he came from, Akhil did not want to go back there. Ajay, the boy from Khusrupur, Bihar, likewise evaded being sent home, not through the obfuscation of name but of place:

> NGO WORKER: When we found him, he used to work at the railway station, sweeping the trains. We took him to the center but he did not want to go back home, he did not tell us his address. Then one day, when we were moving kids from one center to the other, his uncles saw him. They came to me to talk about Ajay. I told them not to meet Ajay because then he might run away from the center. Then we (in the center) talked to Ajay. He did not want to go back home. He said he gets scolded by his father and is not able to converse with his father. When he refused to go home, we kept him in the center. After that we all counseled him and he agreed to go home.

> AJAY'S MOTHER: Even then he was not that willing to come home. I talked to him, I told him that he can leave, but must tell us. He can do at least one phone call and tell us he is okay. We were looking for him like crazy. I told him all this. Then people at the center told him, he can come back anytime if he wants.

What is it exactly that runaway children are evading through such dissimulation, why are they evading it, and how do they do it? Anderson (1991) identifies a kind of enumerative power in the nation's push to map and count. Foucault (1978) famously suggests that discipline fixes, that it is antinomadic. Perhaps, then, in terms derivative of Deleuze and Guattari's

(1987) poetics, Akhil-Aziz is practicing an active and very intentional narrative nomadics, a practice of nonfixity, of flexibility, of fluidity, of never being kept *sur place*. In these ways, then, names may be used to confound the power of the state to arrest movement, to count, to achieve perfect surveillance. The "vagabond" child, in the murky and inchoate space between actively deployed collective names and actively hidden personal names, blurs the categories that render a human being a subject of a certain type and achieves an evasive resistance at the edge of the city's powers, a shadow zone of kinless autonomy far from institutional walls in which selves may be reauthored and refashioned with some considerable freedom.

Petu emerges, then, in such a context and for his part, as a person for whom the expected and expectable variables of locatability and identity, those that become relevant for him only in the constellation produced by his death, are invisible. He thus dies a sort of legal nonperson, kinless, nameless, without the records (or even, of course, the age) so crucial to exist in or under a state (see again Hull 2012, 2003) and later becomes a person or subject by virtue of his subsequent circulation through frames of media and law, but never a citizen.

So much for the immediate official incarnations of Petu's death. Soon I will deal with the legal waves rippling out of it, and the *reactions* of the press, that seem to render Petu suddenly, at least partially, "grievable" and "recognizable." For now, however, I do not want to venture too far—temporally—from the critical event itself, from the ethnographic moment, standing there in the station, at the beginning of the circulation of death's meaning in another sphere: the milieux and channels of the children's informal networks themselves, a sphere where novel meanings accorded to death are formulated and circulate; it is a sphere not readily accessible to the public and political gaze.

Looking upon the mangled body before us, of a dead child, Arshad and Vikram, the two station-dwelling older adolescents who were accompanying us, began to develop and elaborate upon their own theories and narratives to explain the moment (see Nagy 1948; Tamm and Granqvist 1995). In the passage that follows (in which I am J and Vikram is V), Arshad (A) proposes a link between death and such children's precariousness, on

Smoking and scavenging, Hazrat Nizamuddin Railway Station tracks

the one hand, and the economy in which they are situated, on the other: of scavenging bottles (see figures 2 and 3).

J : Do you see this often?

A : Yes, often, two or three per month, not the general public, just children just like this [pointing at Petu]. Not with the general public but with people like this.

J : Why? From the danger of it?

A : From collecting bottles. What happens with the bottles? First I get on the train. From this point they go further along

the tracks, where the train attains full speed, and the children
are arguing about who can get on first. The older boys get on
the train, the younger ones are left behind, and in all this hurry
and hustle of getting on the train, some then fall off and meet
this fate. I'm going to tell you of another incident. A girl named
M——— came here very angry. A bottle-collecting girl. The
Rajdhani was coming on this track, and the girl was standing
right here. She tried twice to get on to the superfast express.
She got on the train once. I was looking from a distance, I swear.
She fell once, and once she succeeded in getting on. I caught
her there. I got hold of her, she got mad at me. I asked why she
ran away, she was all of fifteen.

It is not the economy of waste subsistence alone, however, that Arshad suggests is constitutive of the structures of risk (or the structural violence of
risk) for these children but also the spatialized relations of power in which
the children exist, the train-station life, which emerges as one of only a few
"safe zones" for them, though clearly it is not—neither physically nor in
terms of evading authority—for it is a zone of the state and commerce par
excellence and thus embodies the absence of other options. Some minutes
later, Arshad and Vikram articulate very clearly a further interpretation of
un/grievability and its variability according to frame, subject, and kin(lessness).

J : Don't people grieve and cry?

A : Petu got cut up and died. Everyone is *stamd* [stunned]. But
what are they going to do? There's no time to grieve, the police
are coming and people are afraid and have to run away from
the cops.

How does such a moment travel, circulate as publicly accessible sign
(see Urban 1991 on this notion), if it is to do so at all? In nearly every case I
am looking at here, it does; street children's lives do not seem, as I've suggested, "grievable" in Butler's (2010) sense immediately upon their termi-

nation, at death, but they do travel through frames that offer the possibility of making them so.

Am I, or is this, one of those very frames? Am I helping, through some quantum calculus, to manifest or perform the thing I am describing? That Petu's death had the peculiar feature of being observed by an American anthropologist and an activist researcher was a matter of chance; the whole affair also might have had the possibility of never making it into any public circuit, if it had been covered up or cleaned up quickly enough, if the Railway Protection Force had made it there first. But that is not what happened.

What happened is that both Khushboo and I each went home to our respective families that night and wept. What else does one do after stumbling upon such misery? This is not meant to suggest a romantic or heroic sentimentalism but to inject into this story some of the affective experience that is part of most ethnography. As I have said, I had already struggled a great deal with how to build a bridge between what I had been seeing every day in my field sites and the (emotionally and physically) protected zone in which I was endeavoring to keep my own two-year-old. How could I reconcile them? How could I cross between them without dissonance, and how could I accept the routinization of that crossing? How could I express rage in my daily agitating, as I did (to police officers, for example, or other officials), and then ride the metro home and, say, take a jog in the city park or sit in a posh café? I sought to maintain for my child the very coddled innocence that constitutes a model of childhood whose public promotion I spend year after year critiquing in the classroom. Nonetheless, and despite this, I had just seen and lingered over the body of a dead child, and my son perceived my distress; of that there was no doubt. I imagine he must have heard something, though I cannot remember speaking of it in front of him, for the next day he asked me: what would happen to a child who is sitting on the tracks behind a train that then backs up onto him?

In Khushboo the experience grew into an outrage that took her to the supreme and high courts; she launched Petu into the frames that made his death legible as a certain sort of event, and herself visible as a certain sort of agitator. This would form the "story" the police feared. At that point,

Petu's case began a long process of circulation through courts, institutional spaces, and various media, and he began his transformation into an emblem, a trope, or a synecdoche—a grievable token, in other words, of an ungrievable type, or an embodiment of a proposal that a type be taken as grievable. The question that raises, of course, is whether this transit through the "frames" I propose is a move toward the production of a grievable individual, or only of a grievable *type*.

The petition authored by Khushboo, an advisory consortium of NGOs, and their collective pro bono lawyers was meant to force municipal structures responsible for railway space to implement safeguards and policies for the well-being of children inhabiting those sites, who have long been barred by the law from being allowed to inhabit or move in railway space. To provide them services would be to acknowledge and accept their presence there, and thus the law has built into itself provisions that prevent its own implementation. This approach was to address the immediate cause of danger, but not, of course, the overall architecture of precariousness that *puts* children in danger in general by forcing them into the spatial peripheries of the urban domain, and of labor (including, for example, living in garbage dumps, working with toxic computer waste, or selling sex). The photo was part of nearly every legal proceeding: a "frame," for sure, in transit through many lives, the lives of a death.

The petition presented before the court for Petu's case pointed to a number of legal discrepancies. First, the Juvenile Justice (Care and Protection of Children) Act of 2000 theoretically illegalizes the incarceration of "neglected" and "delinquent" children in conventional prisons or detention centers (though their detention in "remand homes" remains unchallenged). The act simultaneously stipulates that children in "need of care and protection" are required to be presented before the Child Welfare Committee (CWC), and a child in "conflict with law" is produced before the Juvenile Justice Board (JJB). At the same time, the Child Labour (Prohibition and Regulation) Act of 1986 illegalizes the presence of self-employed children working at railway platforms. This makes "railway children" "delinquents" or "criminals" and nulls the requirement to provide "care and protection." Under the Indian Railway Act of 1989, further, children traveling on trains with no tickets can be charged as "delinquents." And finally, under the still-active Indian Penal Code of 1860, a seven-year-old can be charged as an adult, and a twelve-year-old, when in conflict with the law, is always an adult. Some of this may have changed since that time, but it was the legal and epistemological reality that shaped these children's lives at that time.

What all this does, in effect, in a kind of legal impossibility, not unlike the one Beth Povinelli presents in *The Cunning of Recognition* (2002), is guarantee that the only manner by which railway children can receive any benefit or recognition from the law (national and colonial alike) is to interact with the very structures that deny and forbid their existence. The law thus cancels itself out. It becomes nearly impossible for a railway child to gain access to citizenship and legal benefits: the illegality and criminality of her existence make it very difficult to access the very pathway to a machinery of the law that she is required to appear before in order to exist. When their fear of such structures is entered into the equation, the probability that such children will be given any form of due process becomes practically nil. The structure of the law as applied to such children contains so many obstacles to its real application, and so many contradictions, that it emerges neutered. Legal voice for solo children is thus attainable only by virtue of contact with the very structures that refuse to acknowledge them. Recognition under the law is required to provide citizens' rights to children, but law proscribes their presence in this place. The law puts railway children, then, in an impossible situation: it will protect the subjects whose existence it will not recognize. It will require their presence in contexts where their presence is not permitted. Thus it negates its own provisions.

Organizations of advocacy and child welfare could not provide the kind of enforcement that Khushboo was looking for in the prevention of a repetition of what she saw in Petu's death, and indeed, conversely, these organizations would not recognize in Khushboo a legitimate voice: they critiqued her use of the photograph, and one NGO manager even suggested that she was acting "so [i.e., "unreasonably"] emotional" at the sight of "only one" death among so many, such a common occurrence (see Scheper-Hughes 1992 and Scheper-Hughes and Sargent 1999 on this kind of "routinization" of child death). There's an implicit gendered statement there as well. Khushboo decided the only possibility for enforceability was to enter Petu, as a representative of all the others like him, into the formal channels of the law, via the Delhi and the National Commissions for the Protection of Child Rights (DCPCR and NCPCR). Among the first incarnations of Petu in this context was in the letter cited above from the Railway Police to the DCPCR, around a month after he perished:

> It is submitted that, information was received regarding an accident at railway track near Minto Bridge on 4.4.2011. Si Jagmal Singh, PS New Delhi Railway Station was detailed on this information. On reaching spot he found that one boy was run over by a train (Shramjivi Express) and died on the spot. During enquiry it reveals that the name of boy was Rohit age 12 years s/o Shri Gurde Maldar r/o Bihar. There was no permanent address found in Delhi and he was residing as Vagabond at New Delhi Railway Station.

I want to suggest that Petu never attained much public personhood in the months after his death, the period when he might be expected to have done so, as materials set in motion by his death circulated, and that he remains, as one would expect, ungrievable, though perhaps the public notion of grieving for such a *type* underwent a change in this time.

Three weeks after Petu died, an article focusing on railway deaths and the construction of fences to prevent them appeared in the *Mail Today* newspaper. The idea of fences and heightened security to "prevent railway deaths," of course sounds, as does the text above, like it will produce justifications, in the end, for further purging, and in the article the blame is indeed placed on "people who carelessly cross the tracks"; in a similar incident, where a thirty-six-year-old (Bhagwan Parwal) was caught between a moving train and the track and, stuck in position but without fatal injury, bled to death from a fractured hip, Railway Minister Dinesh Trivedi said (in the words of the reporters on the *Times of India* online video) "that the incident itself is sad but that the people themselves are responsible, for [their] breaking [of] rules." In Trivedi's own Hindi-language speech, "discipline" is the focus: "in many places it is seen," he observes, "that in Hindustan there is a lack of discipline," and at length he explains that such things are to be ascribed to disorderly behaviors and are the fault of the people themselves. His explanation packages nicely both discipline as a form of regulatory power (following Foucault 1975) and as a rationalist concept introduced through imperial structures (following Chatterjee 1986 or B. Anderson 1991, "Census, Map, and Museum" chapter). In the *Mail Today* narrative about railway safety, death and risk are features of a generalized and embodied social disorder in the context of a well-meaning authority:

> No death perhaps can be more painful than this. People being cut to pieces under the heavy wheels of an oncoming locomotive at railway crossings. . . . Such deaths are commonplace and many bodies go unidentified and unclaimed. "The children living near railway platforms and tracks are more vulnerable to such accidents," said researcher Khushboo Jain, who works with children on Delhi's railway platforms. Often, kids trying to board moving trains to collect junk or other petty jobs are crushed to death, Jain said, citing the example of an eight-year-old whose body wasn't claimed and an NGO had to finally cremate him. "We take extra care of the underprivileged children in this regard," a senior police officer said. (Irfan 2011)

Khushboo filed a series of legal complaints and, in time, that high court petition, to establish legal measures (not fences) to protect children in railway space. In it, Petu, and his death, are mentioned explicitly several times, thereby circulating the idea of Petu as a person who was initially "loseable," through a sort of Austinian "performative utterance," in elite legal spheres. He is thus, in some sense, manifest, actualized, made real in a public domain, and his death is thereby endowed with entailments. It places demands in the domain of the social and the political (Verdery 1999).

The language of the petition crystallizes as social fact, then, the "reality" of Petu's last minutes:

Petitioner was told by the other boys who had witnessed this incident that Petu was dragged for about 100–150 meters after which his body fell on the side tracks. His body below the waist was cut into pieces. For approximately two hours his body remained at the site and Petitioner, present at the site of the accident, observed the unconcerned attitude of authorities towards this incident. Petitioner also observed that none of the care mechanisms prescribed under the Juvenile Justice (Care and Protection of Children) Act, 2000 (hereinafter referred as JJ Act) existed to the benefit of children arriving and living at the New Delhi Railway stations. For approximately two hours no help from any corner and from any authority reached the scene of the accident.

The text contextualizes the child's death in relations of power and represents it as the embodiment of a troubling element of those relations. Indeed, and perhaps for this reason, it was successful. By the time it had been modified over the course of a year, revised and rewritten, backed by dozens of organizations, connected to several station surveys commissioned expressly for its benefit, and heard in multiple court settings, it had generated its own legal entailments.

Was it Petu who had effected that? The legal fallout from his death appeared in media outlets throughout the country in the days that followed the petition's enshrinement into law. It explained that the public interest litigation (PIL) asserted that "treating children who are living at railway stations as illegal travelers or illegal occupants of railway premises and penalizing them complicates the problem rather than solving it" and "that the central government's (ministry of women and child development, railway ministry) failure to create procedures that protect the safety and dignity of children arriving and staying at railway stations is a violation of their right to life." But Petu had himself, by now, disappeared from the moment (and from narratives of its public meaning).

Petu was not the only such in that ethnographic stint, not the first encounter I had with "vagabond death" in that Delhi winter. Elsewhere, not

Among the Jama Masjid children, near the maidan

three months earlier, two days after my arrival in Delhi, I had come upon another death site that Petu's would recall in more than one way. One January morning, Khushboo got in touch to tell me a child had died at the Jama Masjid (the congregational mosque of Old Delhi, India's largest), and I hurried across town to meet her and find out what I could. When I arrived, I beheld an expanse of ash upon the so-called *maidan*, "the field," perhaps the area of Delhi most heavily populated by the homeless: in tents provided by charities, sometimes, in the winter, and in makeshift huts or outdoors in other seasons.

It emerged that the child had been known only by the name Kaalu (Blackie); nothing else about him (in life) could be ascertained, except that he was a resident of the area and that he was around nine years old. Nobody could relate—or nobody would divulge—any other personal details. Day after day that week we returned, trying to find some snippet of identifiability; we found none until much later, and even that was scant.

This points to a different sort of naming practice, one that is equally prevalent as the dissimulative one I describe above but that might also be more relevant to Petu: of street children known by bodies and *not* conventional or self-fashioned names, by worn selves rather than kin matrices, which is what generate the former. For children whose identity is shaped by separation from kin and place, and who have no property, no dwelling, no documents, and no school, the body emerges as the sole available durable slate on which to fashion and display the self. Sometimes it is the child who authors this embodied self, and sometimes it is an audience, a social milieu; in this case, correspondingly, the body emerges as the most immediate field for the reading of self. And thus, in a field with little displayable or durable capital, what is worn on, revealed by, or achieved with a body becomes absolutely central to the construction of self and status. Where even names and biographies are fluid and contingent, body becomes itself a perduring site for spinning names and corresponding personhood narratives.

Kaalu was purportedly another victim of "safety" lapse, but again also of something much larger, of a position in the world generated by certain histories of subjection. He had gotten a place in one of the large canvas laborers' tents that night, a very cold one, by Delhi standards. Someone had likely tried to light a small fire in the tent, or had dropped a cigarette, and the whole thing was consumed in flames. Everyone got out in time except for Kaalu, who (perhaps by virtue of being a child) burned alive. Just like his identity, his body's remains were also barely discernible in the smoldering embers. By the time I arrived, what was left of him had been cleared, but the other resident children told me that "his bones had been there, just there" in the patch of scorched earth.

In time it emerged that Kaalu was not without discernible kin ties, though they were perhaps tenuous or weak: he was the son of a dealer of narcotic white-out, one of the maidan's primary markets and a substance used (on soaked rags) by most street-dwelling children I have met in India.

"There, just there": Scorched earth, January 2011

Other residents reported that the father had not been located, and they could not say if he had been informed, or if he knew. In the aftermath of the death, the children there were boasting of the loot they'd taken from the carnage: new clothes, new goods. Kaalu's only post-death public itera-tion was in the newspapers, again, which were rather muddled on who he was: in the *Times of India,* he was "10-year-old Kaalu, a beggar from the Jama Masjid area" who got caught in a "death trap"; in the *Hindu,* on the other hand, he was the "boy who has been identified as Kallu, a nine-year-old orphaned boy" who "was burnt alive." A real death of a real person is made public, and situated in the context of failing social services and secu-rity structures, but the person is barely recognizable, nor knowable: he is simultaneously an orphan and a beggar, a nine-year-old and a ten-year-old, Kaalu and Kallu. It matters very little, for these specifications capture his type, which is what seems important for readers to know. He is connected with no kin and with no arrival narrative. His public relevance thus lies in his incarnation as an idea.

In the subsequent months at the mosque's maidan I came to know well the other children living there. One of them was a lank, curly-headed boy whom I will call Shamsuddin: ambitious, proud, clad in a turtleneck, and by his own self-fashioning and self-description, clean (of drugs, and physically). I met him in one of the tents of the maidan, this one run by the NGO Butterflies, adjacent to the one where Kaalu had died; he was one of the boys I spoke to that day about the death, in fact. He enjoyed card gambling, carried around plenty of cash, enjoyed eye makeup and experiments with hair color, and was not without family, though he moved around the area independently. Whenever I appeared with Khushboo, he would ask her time and again if she had any proper job for him. He was known for strong-headedness, got into plenty of fights, and made his share of enemies.

Both of Shamsuddin's parents had been homeless residents at the Jama Masjid, and even an extended family, including an uncle and others, were present. Shamsuddin was associated with an NGO, where he had resided for a short while, and over time, as an example of an exemplary or "model street child," he had been allowed by the charity to serve as a guide on their signature "reality tours." When Shamsuddin's father fell ill with tuberculosis, he sought help from Jamghat, which (by his account) did not or would not help; the father died shortly thereafter, and Shamsuddin fell out with the organization.

Shamsuddin died publicly and with spectacle in the summer of 2012. And in his death and in the multiplicity and contestation of its narratives emerge a new series of tensions and dramas that raise new questions about "street children" and their "grievability," and perhaps complicate the image that is emerging.

Shamsuddin had gotten married over the course of the previous year. It was rumored that he had fathered a child with his new wife, and also that he had remained promiscuous. After his marriage, because his wife was clashing with his mother and was thus blamed for breaking the family, he had moved with her to a different area of the Jama Masjid. After the move, it is told (and again here I am as [or more] concerned with the circulation of story as with discernible "reality"), Shamsuddin became familiar with a girl who was living with another man. He met with this girl daily, and they would converse (and obviously this carries with it a potential imputation of carnality); the story at the Jama Masjid is that she would return to her husband saying that Shamsuddin was harassing her, even stalking her, despite the outward consent she was said to have expressed in her acquaintance's presence. One day, then, the husband appeared as the two were talking and grew so angry that he and Shamsuddin began to quarrel and then fight. Shamsuddin left and paid a visit to his uncle, from whom he requested fifty rupees. It was not granted, as the uncle assumed it would be spent on the pills it was rumored Shamsuddin had been taking. The story continues that the aggrieved husband learned of the loan request and said to Shamsuddin: here, here are fifty rupees, you want to kill yourself? Take it and go kill yourself.

Take it and go kill yourself.

Shamsuddin, now in a rage, left the scene, went somewhere else in the Jama Masjid area, and reappeared in the same spot with a bottle of kerosene. He poured the kerosene over himself and lit a match. The narrators of one version of the story hold that it was not his intention to self-immolate (note echoes of *sati*) but rather that he sought merely to make a spectacle: he lit a match, they claim, and extinguished it, and then he did it again; now it appeared he was trying to put it out, but a strong wind blew, and quickly his nylon clothes caught fire. He tried to take off his clothes, but all too late: the flames were spreading; very quickly the burns took over 80 percent of his body. When the fire was at last extinguished, he was still alive. "Let me die," he is said to have said, "but know that it is at the hands of this man's instigation that I will have killed myself." He remained alive for ten to twelve hours, during which his mother accompanied him to the hospital, where he finally died.

The provenance of the story is itself at issue, not here in my account, but in the legal, institutional, and social realms where establishing the facts of death is always of consequence. It was asserted by witnesses that a police recording had been elicited but that it was deleted. The police said that he never gave the statement, and suspicions arose that they were concealing it. The police, further, wouldn't register any complaint, and Shamsuddin's mother was frantic to charge her son's enemy. The woman to whom this latter was married was now threatening to kill herself. The police intimated to NGO workers that "these people do this every day, and look, see reason, he was a criminal himself." The NGO workers, who had taken up the case and were trying to ameliorate the situation, went to the police commissioner to get a complaint registered; by that time, Shamsuddin's enemy was on the run and was rumored to have paid the police Rs 200,000 (that is, "two *lakh*") to get himself out. Now a new twist emerges, in which the enemy's brother turns out to have been Shamsuddin's best friend. This brother was "booked" by the police in the other's absence, Shamsuddin's mother asked why he should be arrested in his brother's place, and in the end, Shamsuddin's enemy was himself caught and then let out on bail within a few hours.

When Shamsuddin's corpse was at last delivered back home from the coroner to the Jama Masjid, his family and friends initiated the proper Muslim funerary rites (*janazah*), including the bathing (*ghusl*) and the enshrouding (*takfeen*) in a modest white cloth (the *kafan*) (see Metcalf and Huntington 1991 on the politics of handling corpses).

Meanwhile, the NGO that had been involved with Shamsuddin in life and that had been pursuing the case in these late hours, wanted to see some official action, and media coverage; thus, even after the ritual, at their urging, the body was not removed, a violation of proper practice as decreed by Muslim *sunnat*. Bystanders, many of them pious, grew distressed. *Haram*, people said, for this contradicts proper practice. The NGO continued its attempt to publicize the death and, to try to bring in the media, the shroud was removed from Shamsuddin's face, and the fever-pitch fracas, in a process in Delhi not at all foreign to such spaces, worsened. Finally, after perhaps two hours, Shamsuddin's dead body was taken away for cremation. No media came, but his death was nonetheless rendered public spectacle, even fabular in the polysemy and multiplicity of its narration and its entwinement with elements of customary and orthodox faith (see Cohen and Odhambo 1997).

In these moments, Shamsuddin's personhood is somehow transformed. He—the real person who died—disappears, in some sense, in the mêlée of shouts and rumors, as someone with a name; he becomes an aggregation of stories, ideas, opinions, and interests; the *details* of his death become paramount, with the accusations that accompany them; and he becomes an emblem of something. It is notable that it is the revelation of his face that holds the potential, in a community's eyes, to strip him of the dignity that accompanies personhood, and the hiding of the face, by contrast, is the practice tied with proper grieving. And recognition before Allah.

Is he grievable? Maybe.

Is he grieved? Yes.

Does a public mourn him? Know him? Here, far more than in the other cases, the individual or his image lives in death, his death generates its own politics and its own friction, and he becomes an emblem or an embodiment of the imperative of grief. He is situated in kin, he has a name, he has an affiliation, and the conditions of his grieving are themselves the object of contestation. In some ways it is the public politics of Islam and associated notions of martyrdom, injustice, and burial that provide the conditions that make this possible, even though Shamsuddin's death had fewer public iterations than either Kaalu's or Petu's. His face matters.

But what does this tell us? At the very least, it tells us that even for a "vagabond" who exists in the milieu of the homeless the status and proximity of kin is of primary import in the social life of death.

And what about children—do they get to mourn their dead compatriots? Do they think of their own impending deaths? Are they so habituated to death that they develop a tough protective shell in its face? The adult former runaway Amir spoke to this question with me:

> AMIR: The survival rate is very low on the streets. Because there are so many antisocial people that will misuse them [that is, misuse the children], or, just so many things could happen. It's very unpredictable, just I would say that it's like walking in a maze where you don't know, you can't really find a way.

> JONAH: You've told me in the past about friends of yours that died while living there. Do you think, do kids take time to

lament their friends that they've lost or are they not even able to really to have the space and time to do that?

AMIR: Like, you know, a friend died, I definitely had emotions, I definitely had a bond with him but like you know it's not like so much of a bond that it can really affect my daily life, because I'm already used to all the conditions, I've already seen people dying, already seen people suffering, already seen people being addicted, already seen people getting beaten, so those forces come together and make you very strong.

(Un)Deserving Life and Mandatory Loss: Kinlessness, Orphanhood, and the Calculus of Sympathy

The politics of death in the life of Delhi's "vagabond" children is reckoned not just in instances of their own death but also in exogenous judgments of their worthiness that rely on assessments of the presence of death in their own life, wherein their status is evaluated by virtue of an experience of kin death. Death figures into "street children's" lives as an element or indicator of the authenticity of their suffering. In some sense, in other words, or in certain cases, they are seen to have to have been exposed to death, and substantially, to be considered—in some frames—typified, full-fledged street children, the ascription of which renders them, in turn, "unworthy" types. But in assessments of entitlement and charity rooted elsewhere, at the same time, where it is essential that they *be* street children and satisfy some corresponding standard of authenticity (see Povinelli 2002), they need to have an *experience of death* to receive the entitlements that make them deserving. Only orphans need apply.

What are the politics of grief of these dead kin?

Here is a glimpse.

On a hot day late in the months between Kaalu's and Petu's deaths, and well before Shamsuddin's, I stood in a passage beside South Delhi's Kali Temple, also known as Kalkaji, another area popular with (and, according to their own credo, explicitly dedicated to) the homeless, and watched a fierce dispute unfold in a line of street dwellers awaiting their daily portions from a charitable kitchen. The subject of the dispute: the degree to which a certain child should be considered deserving or not deserving of the alms on offer. Most forceful and audible among his detractors' pleas, as still audible in my recording, was a single, resonant assertion: "Uska mami-papa bhi hai!" (But he *even* has parents!). I observed in those months that among street dwellers, and particularly among solo children, some element of death having befallen the narrator, in particular as loss, is such a universal and universally expected marker of the elements of life and lives that it becomes a paradigmatic index of identity and experience.

For these children, then, the precarity induced by a single family death seems a common catalyst for running away; that death is more often than not preventable and induced by some condition connected to poverty; the poverty is, in turn and at some level, often a feature of debt connected to the historical formations I've discussed here. Thus for the Nabil from Bangladesh with whom this book begins (and ends), it is poverty that generates a father's preventable death of kidney failure; money for treatment was not granted ("it wasn't fast," he said, "he died slowly"). And then it is this death that brings in the archetypal wicked stepparent who is so common in these children's narratives (the very day he ran away, says Nabil, his stepfather hit him multiple times).

One path that narrative intersections of kinship and child death takes is an explanation of mortality by recourse to kin absence: if kin had been present, the child would have lived. And it is certainly true. One of Arshad's stories, about another child in a railway accident whose parents' presence was required to carry out the procedures that might have saved his life, if they had been administered earlier, reflects (obliquely) this sensibility: "I told you about that story of another boy who was almost dead *since he did not have his parents.* He had a mother and father [at that time]; even though they weren't willing to come, the doctors still gave him some injections [that is, even though his parents did not come to sign the proper forms]. He was already almost dead. He lived for a few days." Alas, but for kin, is the message; too late, by the time they arrived, to fix this failure of society at the intersection of medicine and law.

I listened—at Jama Masjid, at the railway station, at the Hanuman

Mandir—to an unending chain of arrival and transit stories, mostly from or about children, that tied kin death to misfortune and painful life trajectory. Vinayak, who in a field in Mehrauli near the cremation ghat (the Liberation House: *mukti ghar* or *sansad ghat*) narrated before any other fact about himself the story of his father's death beneath a truckload of quarry stone, and whose mother, whom in time I came to know, narrated before any other detail about her son the fact that he was fatherless. Nizar at the Jama Masjid, who had lost both parents and run away from three uncaring brothers. Raju and Mohan, two Muslim boys who'd adopted Hindu names, who had lost both parents in Jaipur but never mentioned why and fell stone silent when asked. And now emerged a different set of understandings about the connections between kinship, life, and death: one where death figures into the according or reckoning of status.

Trying to make sense of these stories, I began to ask myself: what *does* death warrant? What entitlements are awarded on the lone basis of loss? On the basis of what sort of loss? What can be made of a context where death (as a person's loss or as a life prospect assigned to them with some certainty) is so unremarkable, assumed, expected, present, and unmarked (as is well-attested elsewhere, including, for example, in Scheper-Hughes's [1992] *Death without Weeping*) that *uska mami papa bhi hai* is immediately legible as a qualifier for undeserving status? Is the justification of running away one of the entitlements death warrants?

In the fray of this alley at the Kali Temple, where death's specter and grim promise seem to appear every few feet, can be discerned the clear presupposition that death is and must be a feature of the lives of most sorry cases, that there is something inherent in certain lives that ensures them some experience of death, if not already passed then sure to be coming soon. What can also be discerned is a moment in which entitlements are being denied based on the absence of the presence of death in a child's life. Thus it is not only the expectation of death that is at issue here, but also the expectation of a certain response to death, a certain idea about what it is that the presence of death mandates be done or not done.

Nonetheless, the entitlements are limited, as is the sense of real empathy, or sympathy, or even imagining, of a child's experience of loss or of those lost (as of a lost child), and they reveal some interesting assumptions about who that child is or must be. If in the public life of lost kin the dead are used to assess the status of the living, if loss attaches itself to the bereaved survivor primarily as a social attribute to use in the reckoning of what institutional provisions must be made, then here too the dead never achieve recognizability: from the moment they enter discourse, they rank among the precarious, for whom death is expectable and acceptable. Such children are narrated as having parents, siblings, and other caretakers who (as them) were always already likely to die. In the narrative assignment of entitlement their absence figures in only in general in terms of evaluations of the *need* (or even the burden) for public surrogacy to replace kin: society and its welfare structures should—and should only—need to provide for them (and should only be asked to do so) if and only if they have no other options, only if the other structures that exempt society from these requirements and provisions are, in the last resort, missing.

On Ghosts: Populating the Passage from Life to Death

Dead children leave behind—in narrative, in memory, in geography— traces and specters. Living children, in the presence of so much death, perceive the world around them suffused with such ghosts. Here, in Delhi, where they haunt the landscape and linger long, they serve to help the living make sense of death. They mediate between life and not-life, and they play a leading role in grappling, wrestling, reckoning with death in the realm of the social.

What do the ghost-children tell us about who their cognate selves were and how they were recognized and weighed in life? What can we surmise about the moments and modes of their passing from the discursive eddies and vortices they leave behind at departure? What elements of culture—emergent or older—are reflected in the ghost stories they evince and evoke, and what work do those elements *do* in the world? And, finally, what do children's wider ghost narratives—including about what types of ghosts live where—suggest about their navigations of a death-soaked terrain (see Quesada 1998)?

Ghosts mediate between death and place, which is to say that how a place marks a death and how a death marks a place can be rendered legible or interpretable through the vehicle (or sign vehicle) of a ghost. They also open up a life, after its end, to new forms of authoriality, to reauthoring and narrative bricolage; they allow the living associates of the dead to make much of the kindred soul they've lost, to bestow upon their compatriot (or upon the idea of the dead in general itself, the nonspecific dead) a continuing effectivity, agency, and power that they could not exercise in life. Thus unmarked and unbound they may be recast as empty signifiers, vessels or shells available to be filled with local theories of power and of the social (see Kracke 1988). They allow death to be represented not only as the closing but as the opening of potentiality and of history (or, in some settings, morality and moralized safety: recall Dickens's "The Signal-Man" as well, of course, as his *Christmas Carol*), but they also represent the incarnation, in form, of all fears, the worst endpoint of possibility, the embodiment of the final entailment of structural violence and poverty, even if at times they may also represent the transcendence of such bonds, escape. And they present a warning to the living about the meaning of dying namelessly, unknown and unclaimed. Ghosts are thus for street-dwelling children among the deepest and most potent realizations of the social life of death.

As we look upon Petu's corpse, Vikram and Arshad spin ghost tales. It is not just their content, but their iterativity, spatial and social, in *just this* moment, in just this place, their configuration quite literally in the presence of death, that captures me. If we listen to these two, we see a world in which haunting narrates something about place, about its nature, its dangers, its resident, guest, and chthonic spirits—and what death amounts to in its bounds. Arshad suggests not only that the station is haunted but that it has gradations of ghoulishness:

J : Do the children here believe in ghosts?

A : A lot. I'm an example right in front of you [who believes].
    I'm one. I go to the bathroom. Before I go, I look down to make
    sure no ghosts or any of all that [*bhut-voot*] are going to come
    out of there. First in the toilets I check to see inside whether a

head or a hand or anything like all that [*haath-vaath sar-var*] is
going to come out.[7] And only after that do I sit on it. If I'm my
room alone . . .

So the station is delineated by zones of "purity and danger" (see Douglas
1966) overlaid by binaries of life, death, and the in-between status of the
ghost. Notable that the toilets, sites of filth and pollution, are also where
ghosts are to be found, reaching up toward those forbidden, exposed,
secret, and intimate parts of a vulnerable subject that are themselves somehow outside the realm of everyday culture.

But the village from which the runaway has fled is a different story
(Freed and Freed 1990): there ghosts are more profoundly chthonic; there
they are rooted, not surprisingly, in kin, in the knowledge of elders, in
sacred landscape features like pipal trees, and in starkly different fram-
ings of place; in fact, in the latter part of the exchanges below, Vikram and
Arshad (re)negotiate the potency of specters in their different domains:

> v : One day I was walking in my village. It was late at night.
> As the elders have said you don't turn around and look when
> you are walking at night in case there are ghosts. I ignored that
> injunction and I turned back to check and I got scared scared
> scared. And then I ran and fell down and got up fell down and
> got up fell down and got up [*girte-parte girte-parte girte-parte*].
> I was so scared but I said it must not have been a ghost, I must
> have been imagining things.

> a : You won't believe me but I can only stay somewhere when
> the lights are on, even if I'm alone . . .

> v : I was caught by a ghost for a month or two [see Freed
> and Freed 1990]. When I used to go to school I had to cross
> a jungle. . . . I was caught by a ghost, I went to a tantrik, the
> tantrik would cast me down and hold a thumb and ask ques-
> tions like who are you, where are you from. The tantrik would
> ask the ghost a few questions what is your name this and that
> where are you from this and that who are you this and that.

[This and that=*yeh-voh*=and such.] The ghost said that he is the grandfather of my grandfather and that I stay in their house and in their neighborhood and that I was wronged by them. So then he did the *jhar-foonk* to tie the ghost. There's neem and pipal leaves [and *bhargat,* the triveni tree]. In our village pipal is called Brahma Dev Baba.

But if village ghosts are chthonic and place-bound in the villages from which children have fled, they are certainly not static, or passive. As Freed and Freed observe in *Ghosts: Life and Death in North India* (1993), ghost possessions and hauntings, and surges of ghosts in general, are correlated with shared social stresses (such as might also be correlated with running away, or such as might even reside in triangular relations between ghost possession, social stressors, and running away). Freed and Freed observed, for example, a "ghost epidemic" in 1978 that they correlated with a concurrent sixfold increase in malaria prevalence, a typhoid epidemic, strange weather events, several anomalous deaths, including that of a child, and Indira Gandhi's "Emergency," along with its campaign of forced sterilization (11–12). For the children, critically, ghosts and ghost possession are tied to crises large and small—or the tie between the two, especially insofar as they may see their entire existence as embodying a sort of crisis—and in particular to the loss of what might be understood as an understanding of rationality:

J : Do children run away when they're possessed?

v : Yes, that happens, because when you're possessed you keep moving around.

J : How do you chase the ghost away?

v : When I got possessed I had fever for a few days and nobody could understand what it was so various practitioners [*vaidh vaghera*] came and I started throwing things around, sometimes dishes sometimes clothes so people suggested: show him to some healer, a *hakim* a *vaid* a tantrik a doctor, like that. A snake had bitten me.

J : Do children believe in this in the station?

V : How and why in the station? There are so many people around; these things only happen in rural areas and jungles.

A : Yes yes.

V : Well—if you're alone at night.

A : If you're alone at night won't you be scared?

V : Yes I would be. I won't be here alone at night.

A : You'd be 100 percent scared.

V : They did not kill the snake but laid me on an altar in a holy place and I gained consciousness after some fourteen to sixteen hours. It had bit me on my head and on my leg.

Note the haggling here, in particular, over the presence of ghosts in the city as compared to the village, the critical limen crossed by the runaway child (to correspond to a different critical limen crossed by a ghost). And again, all this is as we are attending Petu, waiting for the police. But it is this final statement on the matter that I think begins to articulate an understanding of what an essentially kinless dead child's ghost would be about:

A : People who die before their time become ghosts.

V : Your time is written [i.e., inscribed] but if you die before that you become a ghost. This is still written in the old Puranas and texts.

The ghost bestows the ungrievable subject a different form of recognizability, a trace that survives even where the person does not in image, media, and archive, a trace that survives *precisely because* its living antecedent should not have died.

The New Hamal: Cast(e)ing Railway Children on Death's Stage

Such understandings of purity and pollution as emerge in these ghost stories — and their reiterations in binaries of space, place, and age — matter, and significantly so, when it comes to assigning children their place *in* death and, more substantially, to assigning them a place in relation to death. The entailments of such cultural conceptions extend quite widely, even into domains of policy, in the ways that children are accorded officially sanctioned roles by municipal, state, or sometimes national bodies.

For whom is it considered proper to handle dead bodies (see Kroeber 1927; Bloch and Parry 1982), and for whom is it considered pollutive, and why? Have such designations changed?

These questions converge on the matter of who it was, in the end, that was called upon to clear Petu's dead body.

I would like to suggest here that cosmologies of purity that inform the answers to such questions are imbricated with emergent *economies* of purity, with which they share a blurry boundary. Thus caste-structured assignations of roles may in fact ultimately be reflected or realized outside the space of caste and in the realm of political economy, poverty, and other axes of identity.

Let me illustrate this.

Once, it was largely or exclusively the dharma by birth of certain castes — the lowest of them — to handle and carry corpses (for an analogue, see Watson 1982). This among all jobs (alongside the handling and treatment of excreta) is considered the most impure and pollutive, a base and vile task to be assigned to and reserved for castes who are themselves, correspondingly, ritually polluted (as denoted more by *aśuci* than by *apavitra*, in Hindu terms). In general, in North India, this fell to the caste group known as Dom or Domba (the term Chandala, now sometimes a pejorative, also denoted specialized handlers of human bodies).[8] It is in fact still the case that interaction with corpses is the domain of certain castes, including Doms, but less exclusively: such configurations are more closely bound to particular places — places where ritual roles are more rigid and where pilgrims are more invested in the correct distribution of those roles (see Turner 1992)—than they once were (the cremation ghats at Varanasi, for example, much more than Delhi).[9]

But the culture of caste laid forth (and seen to be laid forth, and presented as laid forth, following Dirks's [2001] observations of the nexus of caste and empire) in the *Manusmriti* and other of the Dharmashastras has attained a wider distribution than it once had, and a wider potential for applicability, in which it might be said that it is the construct or notion of the existence of a ritual divide between pure and impure, rather than who in particular inhabits its constituent spaces, that perdures, and further that who can be considered an appropriate embodiment of ritual pollution has been generalized—though the process of demarcation persists. In other words, the idea that some people are the proper handlers of the impure and others are not has been maintained, but the occupants of the slots have shifted and the assignment of occupancy to slots has become more fluid.

All of this is to say that in the end it was the other "vagabond" (aawaara) children of the railway station who were called upon by the minor authorities present to handle and convey the fragmentary corpse of their dead compatriot, Petu. It was the aawaara children who were given the job that so long had been given to people whose ritual pollutedness was seen as a feature of their birth, of their place in the cycle of *samsāra* and thus in the cosmos itself. These children, however, are polluted not by caste but by virtue of their position in matrices of age, economy, and history.

This transition is not at all without precedent nor without analog.

Indeed, in Mumbai, children have been issued "hamal cards" in order to license them legally to occupy track space and have been given a particular role in the station to perform just such a task whenever the imperative arises, which is often. Among the stories emerging from our conversations with children in Mumbai was one of a boy treated for severe post-traumatic stress after he was made to clear from the tracks a severed head.

Thus railway children have come to occupy a peculiar place in spaces demarcated by zones of purity and pollution, corresponding to existing differential values assigned to various tasks, but now both the roles and the spaces are assigned their meaning by virtue of postcolonial and global political economies, not formal birth-ascribed statuses, in which it is the unrecognizable living themselves—any of them, regardless, really, of their particular group identity—who are designated the proper handlers of the unrecognizable dead. They assume features of hamal and Dom alike.

What is to be made of this? How might it be tied to the assertions that frame this chapter about the complex calculus of death and subjecthood at the margins of society, about grievability and recognizability there?

Perhaps the hamal cards connect or draw or complete a connection between "vagabond" children and death, as if contact with the pollutive domain of death is properly theirs, as they are themselves already pollutive and polluted, as if dealing with death is their rightful vocation and death is their proper domain, as if their precariousness in life and their intimacy with and constant exposure to death make them just the right people to be dealing with it, as if tradition and modernity have colluded to make them the perfect occupants of death's domain—indeed, as if their predisposition toward dying young makes them the right people to deal with it in life, makes it natural that they should be assigned its menial and final custodial tasks.

Alive, then, it seems, their proper place is in the presence of death, and their proper task is cleaning the space of its traces.

Conclusion: Death's Intimates, Death's Inmates

It begins to come into resolution that children are assigned in publicly accessible narrative a status in death that corresponds to their status in life as subjects more than citizens, as partial people. Children "on the streets" in particular are in such frames incomplete human beings, particularly because of their peculiar relationship to place and, notably, to kin; they cannot avail themselves even of the assets that being situated in a family affords them, nor always do they wish to. Indeed, for such children, even proper names, a vehicle for the articulation of self with kin networks and pasts, are fluid and flexible. The death of such a child is narrated as expectable and thus acceptable, and the precariousness and risk in which the child lives is presented as an inevitable element of a natural social order; the death of the type is lamentable, but the individual is not grievable.

The compelling question that must arise from this is whether or not such children see themselves — their own potential or eventual deaths in childhood, and the real deaths of their peers — as grievable; if, indeed, they see their own disproportionate susceptibility to dying as unreasonable or unjust, or just simply part of the order of things. The account I provide here suggests the latter. When one from among their ranks passes on, he is rarely discussed and barely mourned (expressively); his passage threatens their very livelihood, promises to draw authorities' negative attention to them. Such a loss is occasion for disappearance, for moving on, for avoidance. They expect no different for themselves, should they die. The children inhabiting railway and other public space in Delhi narrate death as a fundamental feature of the space they navigate, of their quotidian experience, of who they are in the world, of an inevitable destiny they will face sooner rather than later (and thus must accept). They see themselves, quite perceptibly but without surprise, and critically but also with resignation, as (more) likely to die. And indeed they are.

Runaway Train

Railway Children and Normative Spatialities

I WANT TO LINGER A LITTLE BIT longer with Petu, that child who died beneath a train traveling to the same region he'd run away from. Petu was a full-time resident of New Delhi's railway station. What does it mean to reside in a railway station? Of the runaways discussed here, perhaps more live in railway space—which constitutes also their means of having run away—than anywhere else. Their place of stasis and stability is found in a domain that is, for others, devoted only to movement.

According to his friends, in the lingo of railway-station inhabitants, Petu "lived at twelve and thirteen" (*woh baara-terah nambar par rehta tha*), meaning his full-time place of habitat, as known to others, was a space between those platforms, where he lived under the "supervision" of another platform dweller, an older youth named Mental. Like most children like him, Petu had a particular place within the microgeographic space of the railway station, a particular social grouping, and a particular persona that had evolved (or that he had fashioned) in and for that space.

Petu's status (even in death) as one of many is centrally important for this chapter. All along these tracks, in fact, and along the others that the Shramjeevi Express was to roll through that day, with Petu on board, are the hundreds of children who have run away from home to live in the stations and even within the trains. They occupy tracks, roofs, adjacent dumps, carriages moving and carriages parked, subterranean passages, and electrical grids. Attached to railway space in India, firm embodiment as it long has been of empire, state, modernity, and capital, are multiple overlapping ideological narratives and investments (see LeFebvre 1974; Massey

1994; Harvey 1985; de Certeau 1984). If such space is ideological, then its use by subjects can also be an object of ideological assessment or commentary, and certain uses can be proscribed, or further deemed defiant, transgressive, or undesirable. A critical task then becomes the determination of the nature of the ideology attached to the space, and the location of the power that determines subversion.

This chapter, then, is about the implications of the active use by Indian "platform children" of railway space in the context of ideologies about (and histories of) that space, ideologies and histories that shape their interaction with it, but not fully. In the shape of such formative forces, children push back, and the chapter is thus also about the way that they are able to exert some meaning-making power over the molding of that space.

Contesting Space: Runaways and Railways

If this chapter is about solo children's use of railway space in North India, it is also necessarily about the way that such use constitutes a violation of regnant norms about what railway space is and should be, and indeed what public space is and should be (and what children's place is in it and the world). The fact is that the runaways whom I refer to as "solo children" violate, on one level, the normative rules of the railway space they inhabit, in part by virtue of beliefs about and histories of that space, and in part by virtue of the nature of their anomalous and anomic visibility, their mode of activity, and beliefs about who they are and why they are there. They claim and occupy zones that are decreed for something else and thereby resist— both symbolically by their defiant presence and literally by their friction with the police—the normative disciplines that govern how space shall be used and by whom.

But in another sense, looking from the vantage of the locus of spatialized normativities, it could be argued that platform children in North India absolutely meet expectations about railway space: they have become part of a landscape and a landscape imaginary, and their presence can be narrated simultaneously as an annoyance and eyesore, on the one hand, and a facet of the sociospatial order of things, on the other. One way of explain-

Body claims: Raju on the tracks, Hazrat Nizamuddin Railway Station

ing this is to suggest that the idea that they are out of place and complaints about their presence have all themselves become normative discursive rituals about railway space.

Nonetheless, solo children's presence in railway stations, on tracks and station structures, and in carriages, is a continual site of visible friction, public debate, and everyday agitation, and the children, in response, push and insist and defy and resist. In this way, railway children's violation of normative spatialities constitutes a true site of violation and contestation, and not one that is simply reproduced in the satisfaction of expectations or rituals. Scheper-Hughes and Hoffman (1998) point to similar dynamics in Brazil: "Street children, by 'invading' the city centers . . . defy the segregated order of the modern city . . . [and] frustrate those who seek to maintain distance and difference from the urban poor. . . . Their claiming of public streets . . . as their own . . . [is] a language of protest, defiance, and refusal. Street children are, in a very real sense, poor kids in revolt, violating social space, 'disrespecting' property . . . and refusing to disappear" (382–83). The resistance of Indian "railway children" has its own special features: it is in the intimate mechanics of their use of space, of tracks, train cars, roofs, in their active possession of virtually the entire domain of the station, with all its components, as a space for the exercise of autonomy and even resistance, that the children's intentionality becomes discernible. I endeavor, then, to document the ways that solo children in North India stake claims, with their bodies, to railway space.

Petu's death, as recounted in the last chapter, provides a further lesson here. Even in death railway-dwelling children (or their bodies) challenge, defy, unsettle ideologies of space. Petu's death, his mangled corpse and all its meanings and entailments and everything that required, made a scene, and a mess to be cleaned up, recalled issues unpleasant to reflect on, drew attention to undesirables not to be shown to the public; as everyone around me seemed to know, a solo child's death (not to mention a solo child alive) is unsightly, and a matter for retribution (upon others whose living visibility might recall the mess of their cognate's death), obfuscation, and silencing. It does not itself sit well, and in recent years such people out of place have more generally not sat well with the new image Delhi has been promoting for itself, occasioned by the Commonwealth Games of 2010, 2020 Olympic hopes, and a new passion for the aesthetics (if not the substance) of the (corporate and shopping mall) modern (see, among others, Srivastava 2007; and Ananya Roy 2003). The matter of such a death—and the presence of other, still-living, children that the death underscores— suggested the police present and, later, public officials, is of a sort that must be closed immediately. Petu's body, living or dead, constituted (following Douglas 1966) "matter out of place," and either way, a nuisance to be cleaned up quickly. But it also constituted a token of a *type* of nuisance against whose appearance a constant and vigilant campaign must be waged if Delhi is to become the type of place that its elites (national government, municipal government, corporate executives) and their subordinates (the police, urban planners, city officials) imagine it to aspire to become.

My first assertion, here, was to say that the simple visible presence of solo children in Indian railway space defies ideologies of railway space rooted in notions of a clean modernity. My second step was to say that such defiance is not just in the eyes of the viewer but that it is also independently real, that it is embodied in interactions between children and authorities. But now I turn to a third question: what, within the contexts and relationships in which it is situated, *makes* it defiance? What is the nature of the power being defied, and of the relations of power in which presence becomes defiance?

Obviously, at the center of these questions are dynamics of the interaction of power and movement, of mobility and motility and countervailing forces focused on fixity. Railway children in India inhabit as full-time residents spaces that are understood as existing for transience and transit only. They embody fixity where ideological structures decree only motion, and motion in a world where fixity is always to be attached to citizen-subjects. Such children thus defy expectations of proper location and locatability (and the proper intersection of age with these other axes), of trackability and staying still when told to (as children are meant to do).

In Foucault's ([1977] 1995) well-known formulation, "one of the primary objects of discipline is to fix; it is an anti-nomadic technique" (218). As a broad observation, this has some relevance for postcolonial railway space, though the relevance is complex. What, for example, might be the meaning of an antinomadic technique in a space that is decidedly nomadic, or at least seems to be? Might it be said that certain fixities can be constituted by forms of mobility as long as they follow the patterns decreed for them, that they are not truly nomadic as long as they maintain the well-worn lines they are meant to follow?

Solo children in Indian railway space, if they navigate an antinomadic power, enact, themselves, an antidisciplinary practice (an "anti-anti-nomadics," perhaps, or an "active nomadics"). They escape fixity, they defy categorical definitives and imperatives, they evade, by virtue of their use of a space of mobility, the confines of processes of subject making to which others are exposed. Nonetheless, as I will show, any resistance or "freedom" achieved must be constantly asserted, maintained, or regained: these subjects are afforded neither the total circumvention of state power nor the empowering perpetual anticentric marginality of which Deleuze and Guattari's (1987) "nomads" are capable. De Certeau (1984) identifies a relevant middle ground between externality and total domination in "multiform, resistance, tricky and stubborn procedures that elude discipline without being outside the field in which it is exercised" (96).

The discipline that is being defied by these children is one that decrees or prescribes the proper nature and form of movement in railway space; the status of the defiant subjects as children, and thus noncitizens, or even nonbeings, of a sort, further charges the symbolic and material struggle over and in this space. Laura Bear (2007) identifies a "railway morality" whose origins in British epistemologies of space "continue to develop within the contemporary bureaucracy" and that has "prospered within the structural innovations of administrative forms of discipline introduced since independence, which built on colonial forms of command" (256; see also Legg 2007). As I suggest in my discussion of the ideological investments of this sort of space in the previous chapter, train cars and stations are in the public imaginary in India usually exclusively functional and instrumental sites of and for movement; runaway children in India seem to see in them, however, a potential to serve as contexts for the kind of antidisciplines I've described, spaces to move to and through: fluid channels rather than barriers, compartments, and boundaries. Indeed, as I will explain below, the train cars themselves are vibrant spaces of passage and transit not to or from any particular station but to and from *other points within themselves.*

The railcars, the objects most readily identifiable as the vehicle only of movement and function, become in this antidisciplinary practice zones with fully constituted internal lives. The Indian train, then, in my view, is *not,* as Marc Augé (2002) suggests for the Paris metro, a "nonplace" (108, 125). In India, railcars *constitute* social spaces, spaces between spaces, liminal nonplaces connecting places to places, but also self-sufficiently constituting bounded zones of being and subject making, of creative self-fashioning, vehicles of movement but also mobile sites of internal movement (Sadana [2010], also refuting Augé, makes a similar argument for the Delhi Metro). De Certeau (1984), in formulating his concept of "railway incarceration" ("Nothing is moving inside or outside the train"), suggests that travelers are "pigeonholed, numbered, and regulated in the grid of the railway car," which forms "a bubble of panoptic and classifying power, a module of imprisonment that makes possible the production of an order, a closed and autonomous insularity" (111). For De Certeau the only movement is in the nexus of window and landscape. But the train in India is indeed a space that is moved *through* (or even *to*), in contrast to the rigid compartmental-ization whose emergence in Europe Wolfgang Schivelbusch (1977) empha-sizes.

The popular image of this notion is certainly not new, nor is it one in narrow circulation; as a kind of paradigmatic myth, it is available in Bolly-wood, and not at all out of reach of the children either who live it out or who are referenced by it. In the "Riding the Rails" segment of *Slumdog Millionaire,* for example, the lead character (Jamal Malik) and his brother (Salim) travel the length and breadth of India as full-time train residents. Ranging the trains' interiors, they sell and steal to survive to the tune of M.I.A.'s "Paper Planes" ("Sometimes I think / sittin' on trains / Every stop I get to, clocking that game"). They ride heroically on the roof, jump from car to car, and, when cast off by a passenger from whom they try to steal food, they find themselves delivered by the journey to what at first seems to be paradise ("Is this heaven?"), but then turns out to be the Taj Mahal itself, emblem par excellence of India.[1] A grand and romantic adventure. It is not at all unreasonable, then, given the flow of such discourses, that many train-bound runaways in India hope for various forms of deliverance like the one captured here.

If railcars can be said to be bounded zones of social activity and move-ment, even as they themselves move, then this is doubly true of the rail-way stations; these are full-time, long-term habitats (one young man under whose wing were several boys in Nizamuddin Station had been there some twelve years since he left his home, at twelve or so, in the Pauri Garhwal Himalaya) for the children who have fled their homes. These too are again *not* "nonplaces," nor, as Schivelbusch (1977) observes of the historical sta-tion in Europe, "alien appendage[s]" (171) to the city. The children who live in these stations ignore, transgress, and defy their lines, barriers, compartments, and boundaries to assert their own modes of movement through them, all the while fashioning their own innovative strategies for circumventing power, hiding, and surviving; enacting their own clandes-tine economies; mounting their own resistance. The stations that I moved through with the "platform kids" during my work were spaces that existed somehow in the shadows of the other, everyday spaces moved through by the customers: parallel (but nearly-isomorphic, coterminous) spaces actively occupied by the children and fashioned to their wishes.

Stations of Life: Rural Runaways' Train-Station Cartographies

All this is best elucidated through an intimate description of the spatial lives of the stations, and of the children's internally shared cartographies of their habitats. As mentioned, Petu, when he fell, was riding a train out of the main New Delhi Station (NDLS) toward Hazrat Nizamuddin, whose specialty is long-haul trains. Having arrived at Nizamuddin, Petu would likely sort through his "take" and trade it for cash either there or back at his home base, or perhaps, having traded them in, he would have collected a new batch at Nizamuddin and then ridden back to New Delhi for a new sale (forty rupees per kilo, approximately and depending on location). While sorting, collecting, selling, or taking a huffing respite at Nizamuddin, he would likely have known and recognized several children there; most children whom I spoke to knew names, faces, or physical markers of children resident in the other stations.

This type of movement highlights the landscape in which the several stations of Delhi, as I've suggested, collectively form a network of movement that spans the city but remains within the bounds of the railway, and a context for interesting forms of solidarity. In a remarkable moment, I watched as a group of new arrivals from Ajmer, some of whom had run away many times and others of whom were novices, met full-time Nizamuddin Station residents and began a session of boasting, sharing information, posturing themselves as more-or-less experienced railway users, showing off their urban expertise and geographic knowledge, and performing what was perhaps a "shared structural predicament and experience of dispossession" (Comaroff 1985): we, they seemed to be saying, are subjects of the same type, on the same type of journey.

Petu never made it to Hazrat Nizamuddin; as I explained, he died at the "Nizamuddin End" of NDLS. The stations indeed have named "ends," and much more; their internal geography, in the children's narratives, is exceptionally detailed. For example, in both Hazrat Nizamuddin and New Delhi Station, the pipal tree forms a central gathering place.[2] At the former it is the primary area for sorting through scavenged bottles and other garbage items, though many of the scavengers that use it, according to the solo children, are residents of the nearby slum Kale Khan (Sarai Kale Khan), and it is seen by at least some as a proprietary territory for people whose primary descriptor is simply that "they have their own homes." The pipal tree at New Delhi Railway Station correspondingly formed a gathering point for runaways, scavengers, and beggars.

The symbolic status of the pipal tree in everyday Hindu (and Buddhist) milieux opens up possibilities for interpreting not only the microgeography but also the cosmography of railway space, for recognizing in what might be taken for a rather banal realm the possibility for a sacred cartography. Certainly sacred structures, shrines and so on, are built into the station's architecture itself, but beyond the fixed structure of the station, trains too can be vehicles of an important component of cosmology and belief: pilgrimage (and, as is invoked in *Slumdog*, the deliverance that might thereby result), a charged and urgent movement of which running away can certainly be an instantiation or an inception.

If the stations have ends, they also have meaning-loaded "sides." These sides usually exist in binary opposition to each other. Different identities are affixed or attached to children in different parts of the station. In some cases, groups from one part either avoid interaction with or pit themselves against groups in other parts; sometimes the conflict is explicit, and at other times it is simply a matter of derision and disapprobation. For example, children who live and work on the Delhi Gate side of New Delhi Station spurn and keep their distance from the "dirty" children who live on the Paharganj side (reputed as it is for prostitution, narcotics trading, and general danger). Douglas's (1966) reading of socially relevant "dirt" and its corresponding binary in "order" is again salient here.

Correspondingly, Hazrat Nizamuddin's Platform 1 (Ek Nambar) kids are seen by the "Platform 6–7" kids, again, as dirty, no good, vagrant drug (over)users with whom they have no interest in interacting, as they say, beyond absolute necessity. Some children, such as Amrit from Platform 1, bridge such zones and are accepted in each. But most find a community, simultaneously mutually protective and hierarchically exploitative, with which they are allied; most such "communities" have an older male who exerts some control, and exacts some tribute (flexible in its form), over the younger ones.

I explained that the first detail we received on Petu from the other children, upon observing his corpse, was that he "lived at twelve and thirteen 'number'" with Mental. Other initial details provided were murky; most of the children present wanted to assert that they had known him but that they were not friends with him, in order to distance themselves from any potential association with the event. Mental was nowhere to be found. The salient element here, however, is the "permanent address," the fixed spot in the station—beyond their connections to particular sides—that most children are tied to. Therefore, in Hazrat Nizamuddin, the primary group with which I worked lived in a nearly invisible spot between two of the inward-sloping segments of corrugated tin roofing that runs along the combined platform, and underneath the footbridge that spans the station, from whose lateral lattice it must be accessed (and falls, according to the children, were quite common, and occasionally deadly). The group, very stable in its composition, slept, ate, huffed, snuggled, and sorted in that location. When I returned to follow up on these children in 2015, the site remained a habitat, but for other children. Moreover, at that time, four years later, "official" track cleaners—with proper t-shirts to denote their enrollment in a social program to give them work—were present alongside the former scavenging children. New Delhi Railway Station had more children on the tracks.

Roof living, indeed, is very common, and a good way to stay out of sight of the police and hostile passengers. Amir, the now-grown runaway who worked at the time of my visit for an organization for street children (and later moved to Atlanta, where he married a Cuban-American woman and received a fair amount of press), recounts that his group lived for years in a spot of roofing in New Delhi Station that they called, simply, "our place." Another group I worked with in the same station lived beneath a

Solitary solidarities: Raju and Dinesh, Platform 6–7 Cooperative, Hazrat Nizamuddin Railway Station

section of stairs near the bathrooms on the Delhi Gate side. The primary
living, sorting, and sleeping location for children in NDLS, however, is the
so-called *jali,* near the Nizamuddin End. A jali is a lattice or screen, but it
also represents seclusion, secrecy, status, and purdah in Indo-Islamic architecture and literature and is thus of great symbolic importance in local collective imaginaries of space; like the pipal, it represents a certain cultural
read of the station. This particular jali, however, happens to be a screen (or a
grill) enclosing an electrical grid or substation (a high-voltage power-supply,
in fact, to feed overhead catenaries); the children's sleeping area, astonish-
ingly, was in the peripheral interior space between the screen and the electrical grid it protected. Despite (my) safety concerns, it seemed to serve its role
as a jali in the domestic and cultural sense, bestowing a degree of seclusion,
separation, security, and privacy within its forbidding perimeters.

During the day, as I have indicated repeatedly, the largely male resident runaways' primary work is what is referred to in local contexts as "rag-
picking." That term was replaced, in nonprofits' lingo, with "recycle scavenging," which has in turn been replaced, to expunge the "scavenging"

Occupied roof space, Hazrat Nizamuddin Railway Station

piece, with "waste recycling" or even "wastepicking." The children themselves tend to use *botal khinchna* (or *khainchna*) to denote the gathering
of bottles into large sacks of woven nylon for later sale. That they do it
and why they do it is central to my consideration, below, of these children's position in contexts of production and consumption; how they do
it and where they do it, however, are of primary importance for the current discussion of antidisciplinary practice, for it is in this activity that they
transgress most markedly the station's normative spatialities and spatialized
moralities, and in this location that they evince the greatest disapprobation.

Collection occurs most often on a recently arrived long-haul train; on
such trains, plastic water bottles are strewn everywhere. Even before all the
passengers have disembarked, the children begin to move, at remarkable

speed, and with exceptionally well-organized division of labor, through the cars. They move swiftly to evade Railway Protection Force sanction, climbing bunks and diving beneath seats, and then move on to another car. After a big collection, when bags are full, and once a lull has arrived, a collective rest period is undertaken to sort and crush. Most frequently, sorting occurs in the midst of those clusters of tracks that have no platform separating them, a space rarely entered by anyone else.

While sitting in the middle of the tracks they are frequently castigated by railway officials and insulted by passengers.[3] Extra food is often a by-product of the collection period. Once collection and bottle crushing are complete the bottles are taken to the stall, just outside the station, of a recycle vendor, who pays the children and then sends the plastic back to the manufacturing facility by way of an intermediary merchant. The co-operative groups of children share the yield, which they use to save to go home, buy inhalants and food, pay off threatening superiors, purchase desired commodities, or send home, on occasion, as remittance. Sometimes the groups are, as I have described, not cooperative but hierarchical and exploitive, and subject to an older-boy boss, as in the case of Petu with Mental. Indeed at times these groups' power structures are embedded in larger power relations. When asked whether the police used to harass him, a former runaway—Anjeel from Roorkee, whose "patron" was Ahmed— explained:

> Why would they bother us? We used to pay them not to bother
> us! They threatened to burn our [scavenged plastic] bottles if
> we didn't. So, Ahmed collected Rs. 20 from everyone, every
> day, and gave it to the police. Sometimes when a higher-up
> police officer used to come, [the station police] wanted us to
> hide. No one bothered me because I started off with Ahmed,
> who was respected there.

In such an account, the line between Ahmed as oppressive gang master and the police is murky; the two serve each other in a neat circle of age-based subjugation, discipline, and control.

On the interior of the station's edges are the designated spaces of the proper and prescribed subjects (or agents) of railway space, including the Railway Protection Force, engineers, train operators, military security (present in force with machine guns), and other station staff. Even menial laborers, whose portering work is formalized in stations, have their own designated location marked with a sign reading "Coolie Shelter"; not so for solo children.[4] There is nowhere they are explicitly allowed to be, or supposed to be, so they must forge of the station's space transgressive zones in which they can move and subsist evasively. This legal and disciplinary conundrum is part of what requires resistance, or transforms just living *into* resistance and defiance of the sort that mounts a challenge to disciplinary spatialities. Everyone legitimate has a "proper place"; even the poorest laborers are legitimate. But for the few subjects who have no proper place, being anywhere is legible as the contestation of a space. To accord them a place in the station would be to acknowledge the legality of child labor and the failure of welfare, to point to the perceived failures of a nation, and to highlight aesthetics that the city and the state, modern, corporate, Asian, and clean as they envision themselves, would rather not dignify by naming it explicitly, or with comment.

Beyond this compartmentalized officiality, and past the tracks' edges, lie the stations' borderlands and shadow zones: dumps, scrubland, makeshift croplands, sewage swamps. At Hazrat Nizamuddin, this zone contains a vacant lot that is called "the Park." While lounging around the inside of an inactive train one afternoon with some eight or nine of the boys, who wanted a solvent-sniffing rest after some tiring scavenging, I asked, after a mention of someone who had gone to the Park, "What happens there?" Raju, with whom I worked extensively at Nizamuddin, looked away and said, in a hushed voice, "*sexy.*" Such forms must also be part of the disciplinary regimes that control children, keep them in place in this space, for here they are a commodity, available in a reliable space, driven to such things by their helplessness. And though such forms are not an official part of the disciplines of such a space, and are explicitly illegal, they exist in a kind of dialectical relationship with the legitimate, of which they are another side. They are, as Nordstrom (2004) describes the myriad illicit forms

Shadow zone: Raju, rushing train, soaked rag, Hazrat Nizamuddin Railway Station

upon which the licit relies, the "shadows"—related to but explicitly disconnected from the everyday, acknowledged, legal disciplines. Perhaps over-
reading his candor and comfort at the question I then later asked if Raju
might repeat what he had said, to open it up for comment by the other boys,
and for Khushboo to hear. But he declined, not surprisingly, and would not
elaborate.

There is also, of course, the World Beyond, beyond even the far
reaches of the station, the part of the city that exists in some complementary relationship to the bounded stations. These are the adjoining bazaars
and slums where the boys trade their bottles (or bodies) for money. Here
are also the service providers for the stations' many denizens: places travelers with a long wait or a canceled train can shop or stay; workshops where
small components for the train station itself, or for trains, can be purchased;
and, most significantly for this chapter, the many nongovernmental organizations serving the children. Children flow in and out of these organizations
and, despite a marked suspicion about the institutions' motives and policies, move between them across the entire space of the city.

In the indeterminate gap between the notion of defiant claim, by the children, and of coercive dislocation, by others, between choice and necessity, there lingers an insistent and nagging question that I have not been able to answer. Is this "their" space, as Amir called it, only because it is the only available space, because street children are cast so consistently out of other people's fixed spaces of habitat that they are relegated to other people's spaces of transience?

An alternative argument to the one I've made above, then, earlier in this section, might be that "antidiscipline" is an over-read, or an erroneous read (or a romantic read: see Abu-Lughod 1990), and that what appears to be resistance in railway space is itself the satisfaction or consummation of the objectives of confinement and discipline; that the invisible, evasive existence I document here is just the manifestation of the place accorded by others to such children, though it is not made explicit, and that what appears as defiance is in fact the push and struggle that such subordination requires of them just to survive in the corner of the world to which they have been relegated, that they are not voluntary inhabitants of the station but rather its prisoners. I do not believe this to be true, nor do I want to.

I recognize that both interpretations can be true, that what appears as "agentivity" from one angle might be choicelessness from another, and that what appears as a space of subjection from one position might be spaces for transgression, or liberation, from another. I recognize, further, that the lack of choice on the structural level—departures rooted in misery and poverty—still allows for the appearance and the sensation, if not the reality, of autonomy and intentionality on another. And fundamentally, the conviction I came away with, as I left Delhi, was that the children fashion and conceive of their lives as the product of an active choice that amounts to a rejection of others' prescriptions; I cannot but take that seriously.

The railway contains its own contradictions, weds and welds together disparate poles of subject making and subjection. In a space whose very definition is imagined to be line and discipline, steel, rock, and timber, antidisciplines indeed emerge. Laura Bear (2007) sees in the establishment of Indian railways in British India the birth of "a new technology for governance" (22). I suggest that this necessarily meant the corresponding and simultaneous birth, at the same moment, of a new space, malleable in its nature, for *countergovernance* (and perhaps antigovernmentality). As the power operant or embodied in the railway shifted from empire to state, the mode of using its resources in the service of resisting power must further also have changed. The genesis of the railway provided the preconditions for its own rejection and resistance. And not just resistance *to* itself, but a wider resistance *born by virtue of its existence*. In the laying of the tracks, the seeds were planted for the deployment of railways as both instruments of power and also themselves as vehicles of defiance.

Embodiment

Children mark railway space in India, but the railway also marks children; their interaction with it is embodied, inhabited. They wear their intimacy with the railway and its disciplines on their selves, and on the outside as well as the inside. The somatic trace it leaves (not to mention the emotional) is indelible and permanent. Thus a labor of empire and nation and industry becomes a feature of the solo child's subjecthood, part of the landscape of the body, just as the solo child becomes a feature of the landscape forged of that material assemblage.

Recall now Nabil, the thirteen-year-old runaway from northwest Bangladesh, who lost his arm, as he slept one day by a train he thought inactive, to the space between a train's wheel and the tracks; he now cannot bring himself to go home for fear of the grief he is sure he will bring his mother by virtue of his railway-mutilated body. Recall also that Nabil, in the space of "the street," was Toonda, a nickname he has been assigned that means "cripple" or "one-armed." During his tenure at Hanuman Mandir, above which the trains lumber into Old Delhi Station, he could not be located by any other name. Thus his reshaping by means of the railway here becomes identity and persona, and the railway becomes in his narrative of self life history. Petu and Nabil challenge us to see, then, one of the ways that it is possible for histories of empire and industry to become life and death and embodied subjecthood by means of the railway.

Discipline, observes Foucault (1975), is a matter of the governing of bodies through the vehicle of naturalized norms suffused into the nearly invisible fibers of daily living and the subject's concomitant submission of her own body to rules of governance. In previous sections I have underscored the ways the railway is tied up with the body's governance through disciplinary mechanisms. What I am pointing to here is something rather different from discipline; it is about the materiality of interaction with a historical form tied to capital, migration, and rule, and the way that materiality inflects lives and produces subjects and categories. Petu and Nabil reveal the intimate and nearly limitless reach of the postcolonial railway into the domain of the body, subjective being, life, livelihood, emotion, and death. In coming sections I will deal with the extent of the railway's reach in a different sense, with its tentacular spatial reach into remote rural lives, with the yoking of such places to other places and processes through specific historical acts.

Liminality or Mobility? Tracks as Sites and Rites of Passage

Kovats-Bernat (2006) proposes that street children in Haiti occupy a liminal position, in the Turnerian (1969) sense, and that they are further stuck, suspended, in a kind of rite of passage, but with no destination or exit. He suggests that they embody

> an interrupted rite of passage whereby domiciled children are transformed into a liminal state where they remain and are generally vilified variously as aberrations, monstrosities, or social threats, and in all cases as nonpersons. . . . [The street is] an unstructured space of seclusion and marginalization where liminal identities have been assumed after old ones have been shed, without any promise of a full reaggregation back into structured society with the status, rights, and privileges that come with being a full Haitian "person." Is the street then the site of a ritus ruptus, a ritual of passage interrupted at the *limen?* (69)

Could the social landscape I have described above be well explained by virtue of liminality instead of discipline and evasion? As part of a larger cohesive social landscape, less turbulent, less dissonant, less disjunctive?

Certainly it is true that railway children appear suspended, and that what they are suspended in suggests the idea of passage from one place or state to another.

But a liminality argument, by claiming that street children occupy a space that is essentially and immutably in-between, presupposes a certain top-down force exerted by social space over such subjects and ignores subjects' ability to exert, to claim, to control, as imagined in a Lefebvrian (1974) sort of contested space, or the ability to use, mold, and interpret, as we derive from de Certeau's (1984) conception of everyday space.

Liminality as a problematic for this contains presuppositions about who sets the dominant rules of space (always distant and elder others from a unitary "society"). The notion of "re-aggregation into structured society" assumes that "street children" are not already part of the social fabric; hints at an equally unsatisfying corollary that suggests that there is indeed a unitary and uninterrupted social fabric that they might otherwise be participating in, and that everyone must be either inside or outside of its bounds; and favors certain axes, sites, or agents of aggregation while denying others. Finally, a liminality argument suggests a permanent nonarrival and, failing to accord subjective perspective what it deserves, ignores the possibility of interpretations of permanence and stability in the space deemed by other observers liminal, or, alternatively, the possibility of children's locations in railway space as always allowing for a thousand possible and impending arrivals.

Mobility, as an idea, problematic, or line of inquiry, offers a much richer framework for reading railway lives than liminality (see Urry 2007; Cresswell 2006; Vannini 2009; and Greenblatt 2010, for more on mobility as paradigm), helps explain the distinctive allure of railway space and account for what it offers. It highlights questions about what certain technologies and spatialities offer children, and what active use they make of them, rather than imagining them as hapless and helpless failed beneficiaries of a social status that might have been bestowed upon them.

More specifically, in interrogating and interpreting the significance of the choice of this particular location (if it is indeed a choice and not a product), I suggest railway children endeavor to ensure that they are never severed from the experience of escape, of running, that they seek to be permanently fixed in the space of movement, of nonfixity. Thus they are always, in some sense, able to remain in the state and process of running away from anything that threatens to become a home, they rarely leave the conveyance that delivered them to their current location, and they remain always able to be anywhere they wish with a sprint, a jump, and a hoist. In such a framework, they are maximally mobile, and what this mobility offers is possibility, the possibility of the perpetual availability of movement; it becomes a potential mode of empowerment, or rather a mode of perpetual potential empowerment, and provides an exceptional opportunity for the exertion of control over self and body (in a very limited menu of such sites).

It is a significant fact, for many reasons, that the "railway children" I work with exploit the ability of the trains, with great frequency, to move across the full breadth of the expanse of the subcontinent, to serve as a vehicle for deliverance (not entirely unlike in the model suggested by the *Slumdog* scene I mention earlier, where deliverance is a sudden magical arrival simultaneously at the Taj Mahal and at pubescence). Departures of this sort can be on a whim, and without preparation, and, new runnings away as they are, they are often precipitated by a narrative of confinement or boredom, or by troublesome entanglements demanding flight. They are also another mechanism for the evasion of power; they are frequently a response to a solo child's sense that too much is known about his or her whereabouts, or that too much time has been spent in a place. Some such peregrinations can be seasonal, aimed at escaping heat by traveling to the hills; others might be economic, driven by the potential gain to be had by begging or stealing in a new place, or at a festival.

Amrit, for example, periodic resident of Hazrat Nizamuddin's Platform 1, boasted that he had been all over India on his own, to Calcutta, to Goa, to Kerala, to Bombay, ever since leaving Madhya Pradesh to escape his violent father; whenever the desire strikes him, proud as he is of his perpetual freedom from ties, he sets off down the tracks, evading ticket collectors and the demands of station friends alike. On one occasion, as I sat with a group inside an inactive train car, I said, jokingly, "Let's go to Goa. Right now." I was unprepared, and surprised, when my companions sought

Sprint/jump/hoist, April 2011. Note the child clown in the background of bottom photo. The train is always moving during this process.

to clarify whether I was serious, suggested it might be a good idea, began clamoring excitedly about the prospect, and asked which train we should leave on. No, I had to say, not really, just kidding.

Departures to the city and from home in the village also may start like this, as an assertion of the right—or bravado—to leave, even on a whim. The idea that trains offer perpetual potential possibility through the mechanism of mobility starts at home, for that is how the children arrive in the city, where they continue to use trains in this way, to begin with. If urban railway space is "our space," in their view, rural platforms are equally so, if not more. Runaways recount that it is the presence in their village of stations and platforms that long offered them a promise of escape that they ultimately accepted. Elsewhere, however, the sole obstacle—and key to—running away is *getting to* the station, and further also potentially avoiding the closest station. And yet, in all such cases, railway space is central to the narrative, indispensable to the process. Here is how one informant explained it:

Q : And you went by train.

A : By train, yeah, it's hard to, you know, we traveled a lot, we walked a lot to get to the train, because, at the time, my village didn't have very good station, in order to go to the big cities like Mumbai, Delhi, we had to walk about like more than a half day just to get to the station and then we hop onto the train.

Q : So did you go frequently to . . . the station was not in K——?

A : No, at that time only local train used to go there but since we knew that we were running away we didn't want anyone to see us at that local station and also we had some money with us because we stole from our home, and like they could catch us, so that's why we, we walked.

Q : And then what happened when you go to the larger town, with the train station? You got on the train, and you bought a ticket for Delhi?

A : Oh, no no, we never bought tickets. We went there without
tickets. When we went to Delhi it was probably crazy, like, because we never seen such a thing in our lives, so we were just
astounded by just like looking things, so we just, we had no idea
where we were, so many people, lines, noise, like people are
like, cops are in uniform, it was just something that I had never
experienced and expected to see and then when I think about
it, I think about that station, I mean it was totally different, like,
I mean it was astonishing, it was a really happy moment, I don't
know, but that was just something very different in life and I can
never forget that.

In sum, my intent is to propose trains not only as a site of resistance, as
I suggested above, but indeed themselves as enablers of a resistance—and
a transformation—that is at its core situated in mobility (and whose effectivity is derived from its objects of opposition, the various forms of power
embedded in fixed places, or enacting fixity); among the fruits of this resistance are untrackability in the context of real and perceived surveillance,
even by families or peer groups, and circumvention or evasion of spatialized
disciplines of city, state, and labor. In all this, it is thus something rooted
in the character of railway space itself—movement—that offers something
special, and that explains the primacy and unparalleled promise of trains
in this constellation of spatialized relationships. At all costs, in individuals whose whole identity is tied to histories and narratives of mobility, the
ability to move at any time, the ability not to be confined and to maintain
autonomy, is valued above all and must be maintained. It is thus without a
trace of surprise or affect that children note that a longtime resident of a certain place in a certain station, suddenly unlocatable, may have "moved on"
to Mumbai, or Haridwar, or somewhere altogether unknowable.

History as Intimate Experience:
Leaving Home in the Postcolony

The runaway's transits and pathways to the city from the village manifest
historical forces—capital, nation, empire—in individual lives. Even the

specific contours of each runaway's passage to the city bear the traces of the past and the imprint of the global. Children leave without permission or companion from their villages, and unwittingly, as they travel, they follow the precise pathways laid by industry and history; the changes wrought by those same forces, with their tentacular reach, are the antecedents of their departures. From the frame of the "history of the present," running away is unambiguously a historical act. In the decision to depart are concentrated the elements of three centuries of historical transformations. Ghosts of more than one variety haunt the tracks. The train becomes, for the potential runaway or real runaway, a catalyst able to transform a childhood into what is interpreted by observers as a "street childhood," and it thus becomes a crucial node in the relationship between rural people and larger markets, between children and the city, and, even if only for separation or reunion, between children and the families they leave.

Most runaway children in North India, as already established, flee from a set of shared rural predicaments generated by interwoven histories of industry, empire, and urban growth, or rather they flee from emotional and domestic predicaments inflected by these historical formations. Some of them are the gradual runaways I've described, and others are accidental (or even casual) runaways. But the key is that their literal vehicle of departure is a product of the same histories, moments, and forces that have generated the departures themselves. It is in this context that the "railway children" and the railways themselves embody (passively) and enact (productively) a certain historically significant relationship between city and country (and not to mention, as I will address below, between citizen and country, in the latter's other sense). My purpose in this section is to explore the tie that the railway cements or suggests between the rural and the urban, and further to develop an understanding of the human being that resides at—or indeed lives—these interstitial crossings and junctions: the runaway child who is conducted by virtue of tracks laid by empire, capital, and the postcolonial state from village to city.

Poor, railway-dense states like Bihar have stations in or near almost every village; density of railway networks and frequency of railway movement in and out of a village ends up a strong predictor of running away. It is further at a very young age that many families socialize their children to the rite of working and selling goods like tea, refilled water bottles, and trinkets in such stations. Their exposure to railway space begins very early, then, and the notion of its potential for movement is easily naturalized. The web of railway economies in fact extends right into the village, where gangs further recruit children to work in the stations. And, as villagers have characterized it, children move slowly away from home and into railway space; thus even running away may be a process more gradual than one might assume. One returnee from Uttar Pradesh, some six weeks after his repatriation, describes his running-away—indeed more of a moving-away—as follows:

> I was about 11 years old when I ran away. I was frustrated that I had to work—fetching water again and again. I didn't really think that I was running away, I just did it. I was in a habit of running away from home since when I was very little. I left Meerut, took a train. I asked around and realized I had reached Kota. There I stayed for a month, ate food by selling water bottles in the railway station. Then I reached Delhi, where Salam Baalak trust found me. I spent years there; now I've come home.

Another returnee, Roshan, described his transit in a similar way: "I tried Muzaffarpur first, and I begged there. But they catch kids in Muzaffarpur. So I ran away to Sevalpatti. Stayed there for two days, begged again and finally set off to Delhi." It could well also have been the case, for example, that the children's informal railway-station labor depicted in *Lion* was more of a regular component of the Khan family's life than the film reveals; the exception might not have been Guddu's visit to the station, but rather what went wrong on that one visit.

The passage of one now-grown runaway from Bihar to Delhi reveals the profound extent of the railway's reach into the intimacies of selfhood. The respondent's life story is inextricably imbricated with trains, even from birth. His father, he claims, was a railway employee in Bhagalpur, an area from which an exceptionally high number of runaways depart. At eight years of age, he became a sort of "accidental" runaway by virtue simply of boarding a train to Delhi, and for eight years he had no further contact with his family while he lived in the city's railway station. In a remarkable further elaboration of his relationship with the railway, he later directed for a nonprofit organization a "reality tour" of Delhi's railway station—and he worked with new child arrivals from the countryside.

Though he claims that life in the village "was good . . . not even that bad" and though "there wasn't too much poverty" there compared to other villages, he chose the Delhi railway station over home. He further didn't "come" to Delhi, he insists. That would be too intentional a characterization:

> I was taken.
>
> I was with my friends. I was excited to see Delhi, to know new things. Delhi is different from the countryside, you know. People in my village always talked about how great Delhi was. So I said: I'm going.
>
> I did not come on my own. I had some friends, they had some railway tickets. They wanted to go for vacation, for eleven days. Come to Delhi, they said, Delhi is great, there are strange things there, there's a lot of electricity there [that is, "it is an electric sort of place"], so I became very excited. I came with my friends, and then they abandoned me [on the train]. I had no idea what Delhi was, I was totally senseless [that is, bewildered]. So when the train stopped, some people said we're getting to Delhi, it's the last stop. And I got off.
>
> I saw how different everything was, what my nation is like. On my first night I slept on the platform. In the morning I met some other kids. I was hungry. I saw some of them eating. You see, they noticed me because I was well dressed when I first came.
>
> And I lived in the station two years.

This interview sheds some light on the process by which railways mold lives in places like Bhagalpur and continue to do so in Delhi, sometimes even after "street childhood" is abandoned.

Nation's Capital(ism): Labor, Production,
and the Runaway Train

The Indian railway system was built under British administration of the subcontinent, as a joint-stock venture, in particular in the expansion of an industrial infrastructure designed to profit English commerce (see Kerr 1995; Andrew 1884; Tiwari 1941; Thorner 1950). Under the independent postcolonial state, trains remained a most basic component of efforts at nationalization, including both symbolic efforts at "integration" and strategies of accumulation. The imperial structure of trains and tracks themselves (see Kerr 2007; Aguiar 2011; Bear 2007) are imbricated in complex ways with the rending of families by other imperial forces. Dirks (2001) sees the history of the "steadily increasing economic investment in imperial power (propelled in particular through the joint stock arrangement of the railways and other infrastructural projects)" (43) as a most basic engine in the transformation (through its extension into the domain of culture) of colonial rule. The train delivers such mobile and marginal subjects as runaways to the very cities whose growth generates the anomic conditions that mandate their departure. They move along the lines themselves laid by labor, and along which flow labor, but these children exist at some level exterior to that labor; discerning their role and place in systems of production is complex indeed.

On the broadest level, clearly, postcolonial South Asian railways are a space associated with but not confined to labor.[5] They are of course themselves a product, in their physical existence, of human labor. But more generally they are also a product, if not a precondition, of industry and manufacture; at the very least they serve its requirements and exigencies. They are, further, a generative producer of labor, and they encourage and allow participation in dominant systems of production. They thus constitute an assemblage that sits in a complicated and orthogonal but complementary and intimate relation to capital.

And despite all this, for "platform kids" the railways' relationship to capital means something rather different; in the local genealogies of their departures from home capital has played a significant role, and indeed both the *presence* of trains in their villages and the *place* of trains in their villages have everything to do with relationships between systems of production and local life. Village children of the sort that end up alone in the city could certainly be said to be doubly disenfranchised, as poor people first but also as children, by capitalist relations. But runaways are not part of even the very same labor flows that have strained and stressed local lifeways and their own families; rather, though they use its fibers and ride its drafts, they are a by-product of those constellations, embodiments not only of material constraints but of emotional reactions to material constraints.

But it is the plastic bottles, perhaps, that reveal the most, provide the greatest window on these subjects' position in systems of production and consumption: consumer comforts, provided by Nestlé or Pepsi, emptied, rendered shells or skeletons, ghostly refuse of completed train journeys, and translated into their own forms of subsistence for people on the very edge (by virtue of their age, their location, their social status in the village, and their subsequent or consequent social status in the city). What is to be made of collecting bottles in a postcolonial and global capitalism? Scavenging illuminates well, as material reality and metaphor alike, the peculiar relationship between these children and histories of relations of production—a window onto the meaning of the children's experiences in a larger economy.

Scavenging of plastic bottles is an activity that follows acts of consumption (in its most basic form) and makes life from waste. It is an activity that relies on vessels emptied by the wealthier users of the trains. It is an activity that happens on the very peripheries of the stations (and of economies), in the forbidden zones: between tracks, under trains, in borderland dumps and emptied and closed cars (see Chakrabarty 1992). It is an activity that constitutes a system that survives only because of children, and only because of their marginality. And it is an activity that is made possible by virtue of a quasi-commodity (water bottles), void of the content that gave it its original value, available only in late capitalist modes of production, legible only as an idea (and as an object of desire) by virtue of globally circulating norms, and recyclable only by individuals situated in postcolonial modes of subjection.

Burdens and payloads: Scavenging bottles, March 2011

Signals and Semaphores: Reading Ideology in Railway Space

It is in relationship to these formations and aggregations (of power and production and their intersections to each other) that railway space is endowed
with ideological value and thus becomes a charged space, and an object of
contestation and struggle, somatic and semiotic and material alike. But with
what kind of ideological formation is railway space invested? Why and how
is it charged, and who is attached to what its meanings are or should be?

I started this chapter by asserting that runaway children's physical
location is legible only as violation or defiance in relation to some ideological formation that decrees or prescribes what railway space should be

and what children should be, makes assignments of who should be where. I suggested that the power relations that shape children's lives in postcolonial railway space emerge from the ideological valuation of the space in question by observers, from the simple presence of aberrant and anomalous subjects who are seen to violate that ideology, and from the active rejection of ideological prescriptions by the children; ironically, perhaps, it is the same ideological framework that assigns *both* children and railway space their imagined value, expects them to be in the same space, and then denies that possibility.

The ideological imaginary of the railway is itself tied to its location in the matrix of nation and capital: railways were the great and heroic engine, so to speak, or vehicle, or locomotive, of Progress, Production, National Unity, Cohesion. Railway space, most generally, is assessed ideologically according to notions firmly established in that framework or set of frameworks to which the label "modernity" has been affixed (see Todd Presner's [2007] *Mobile Modernity: Germans, Jews, Trains* for a compelling account of the entanglements of railways with visions of nation and modernity). Railway space in India is supposed to be the exclamatory, triumphant (even if ultimately triumphalist) incarnation of the truth that is invested in the modern; on that steed, envisioned the colonists and their allies and heirs, India would be carried forward to a new and glorious age. The heirs tempered their visions, perhaps, as the more mundane exigencies of administration and corporate production took over from the grand projects of empire and early nation, and in that moment railways began to become what they remain: the elementary filament of daily life, as in Amir's history. If the railway is an ideology, then, it is one that has constructed itself around the idea of the modern.

David Harvey (1998) has suggested that high modernity was/is characterized by a "belief 'in linear progress, absolute truths, and rational planning of ideal social orders' under standardized conditions of knowledge" (35) and that high modern*ism* was associated with a "surface worship of the efficient machine as a sufficient myth to embody all human aspirations" (36). De Certeau (1984) evokes a similar sentiment in his own distinctive way, calling the city "simultaneously the machinery and the hero of modernity" (95). I suggest here that the railway stations, trains, and other spaces whose rules the children defy by their presence represent an embodiment of some larger form of knowledge if they are not self-sufficiently a form of knowledge themselves. The implication then becomes that runaway children in railway space are seen to constitute a violation of proper modernity, and of its aesthetics.

Poignant iterations of such metadiscourses of railway modernity are written right onto the walls of the city's stations and train sidings, inflecting the space with an elaborate commentary on itself, telling readers how they are supposed to be interpreting its meaning and purpose. Sweeping across the side of one train, in multiple pieces (and in other iterations on other cars) and swallowing even windows and doors with its digital mural, stretches an image of a ravishing palm-fringed golden beach, at dawn, a few bathers in the cerulean water. The image, in front of which the children sort their bottles and sniff white-out, reads "Go Kerala: God's Own Country" (drawing now dark and certainly accidental resonance with Arundhati Roy's [1997] widely accessible tropes-upon-tropes, for instance of "God's Own Country," in *The God of Small Things*). The train is once more incarnate, or tries to be, as a distinctly luxury-class vehicle of deliverance (and advertising). The side of another train, the words punctuated with photos of thrillingly happy middle-class lives lived by fair-skinned models, reads "Life Plans / Growth Plans / Health Plans / Pension Plans / Child Plans: Get More Out of Life. A Plan for Every Need." A third bears a mural, sponsored by the venture "Max New York Life Shiksha Plus II," showing a manic (also fair-skinned) teenager, jumping, surrounded by bourgeois teenagery sorts of things (electric guitars, basketballs), hemmed in by giant textbooks (economics, biology, math, English, physics). The text reads, "Child Plan for Education. Plus Talent." Again, directly beneath this visual insistence on what sort of truth rolls across the tracks, as with the others, the railway children sift, sort, sniff.

Nearby, at the National Railway Museum, a remarkable space of alternate imaginaries and faded utopias, lie further (and anxiously uncertain) elaborations of an ideology of what trains are and should be. On a trip with my two-year-old, I ride a minitrain, its locomotive in the imperial style and its passenger cars bedecked with Mickey Mouse, Donald Duck, and the rest

Life plans/child plans/plastic bottles, Hazrat Nizamuddin Railway Station

of the Disney cast of characters, which travels, as if a time machine, through a fantastical landscape encompassing all of India's railway history. On the slow circuit, which includes a real "engineer" and a ride through a mock tunnel, we pass viceregal sleeping cars, workhouses of early industry, oddities of local experimentation, and the extinct dinosaurs of everydays past that are perhaps not so terribly far away as they seem. For many children, *this* is the Indian railway: an image, an idea, a trivial fantasy best kept in the realm of fun, a checked box in the logbook of modernity's completed accomplishments. An object of consumption, play, and pleasure.

At the visitor center, near where the toy train leaves from its "station," the gothic title text of a large interpretive placard reads "India's Rail Heritage," the letters in the colors of the national flag. A railway map overlaid with icons and what appear to be 1970s stock-photo bubbles suggests some ways that the railway serves as a unifying filament, tying together a landscape of difficult diversity. On one side lies a list of "landmarks," a timeline equating progress with the progress of the railways. And in the lower corner a metatext explains:

INDIA'S REMARKABLE RAIL HERITAGE, UNIQUE ROLLING
STOCK, CONSTRUCTIONS AND MANY WORKS OF GENIUS MAKE
THE INDIAN RAILWAYS A UNIQUE EMBLEM OF CONVICTION
AND INNOVATION SPANNING ABOUT 152 YEARS. DURING THESE
YEARS, THE INDIAN RAILWAYS HAVE SERVED AS A LIFE LINE
FOR ECONOMIC, SOCIAL CULTURAL & TECHNOLOGICAL PROG-
RESS. FROM ITS HUMBLE BEGINNINGS IN APRIL 1853, TODAY
INDIAN RAILWAYS DIVIDED IN 16 ZONES & SPAN 63,122 ROUTE
KMS., CONTINUE TO SERVE AS A LIFELINE OF PROGRESS UNDER
SINGLE MANAGEMENT.

From Station to Nation:
Railway Children and the Problem of Citizenship

But where might one find that "single management"?

At the top right corner of the placard in the rail museum is an emblem
of the national railways that reads:

IN THE SERVICE OF THE STATE.

Back in the stations, along the grittier and smellier living tracks of the real
railway, the truth of this is perfectly visible: there, indeed, the runaways'
zones of movement are patrolled by agents of the state: the Railway Protection Force, the ever-present elite military police, beautifiers or modernizers
from Delhi's municipal commissions, functionaries of children's homes.
The same agents who were so annoyed by Petu's body and our questions.
And as with so many things, in the railways the state is the heir of empire
(see Chatterjee 1985, 1993; Dirks 2001), and the place of the railway in the
triangular constellation of production and polity, though altered, bears an
intimate relationship to the past.

And for the children too the railway serves as a line of subjective connection to nation and state, a hitch between their mobile lives and their
(self-designated) anomic homes, on the one hand, and the unruly formation
that is India, on the other. As the railway museum placard suggests, and as

Dinesh encounters a railway employee, March 2011

Amrit demonstrates in asserting his ability to leave at any moment for Calcutta or Goa or the Himalayas, just as he originally left his father's unkind
hand in a distant district of Madhya Pradesh for Delhi, for runaways and
other classes of child migrant-mendicants, whether on inceptive or intermediate journeys, the railway serves as a unifying filament for the entire
nation. Whenever I wish, whenever a breeze of freedom (as they call it)
touches me (and what is freedom, here?), they say, I can leave: watch me.
They ride its length and breadth, thus delimiting its very shape with the circulation of their bodies, manifesting and performing its existence with their
pilgrimages and journeys. And if the postcolonial runaways mark the shape
of the nation with their rail journeys, in turn and inversely they also mark

their own narratives with the shape of the nation. The cartography of their own longings is circumscribed by it just as the cartography of the nation is circumscribed by them.

Not, of course, a new notion, all of this. On the idea of rail as nation, the colonial-era Indian statesman Madhav Rao wrote in 1885:

> What a glorious change the railway has made in old and long-neglected India! . . . In passing from the banks of the Tambra-purny to those of the Ganges, what varied scenes, what successive nationalities and languages flit across the view! Tamil, Telugu, Canarese, Marathi, Guzerati, Hindustani, Bengali — populations which had been isolated for unmeasured ages, now easily mingle in civilized confusion. In my varied long journeys it has repeatedly struck me that if India is to become a homogeneous nation, and is ever to achieve solidarity, it must be by means of the railways as a means of transport. (quoted in Kerr 2007: 4–5)

Recall that retrospective assessment by a former child runaway, 125 years after Rao's journey, of his arrival in Delhi: "I saw how different everything was, what my nation is like."

If railways are at least in part spaces of the state (just as they were once a space of empire and just as they are simultaneously a part of global systems), then a question of citizenship arises. Citizenship entails subjecthood, of course, but it also includes new elements of membership and rights less present, or differently articulated, under colonial administration (and rights that the interaction of globalization with sovereignty and capital is changing) (see Comaroff and Comaroff 2006). If railway space is now an interstitial point of contact between solo children and the state, what is the status in that encounter of the subject within the polity? I have said that this railway space is a site of governance and governmentality and their resistances; if this is the case then the state is primary among the assemblages doing the governing, and the children are thus subjects of the state in particular, and more than elsewhere, in this domain. They are governed in it. While they mount, as I have suggested, a substantial resistance, they also acknowledge it as the power operant in their living space, and, as is the case with all hegemonies, they are on the broadest level compliant. Even evasion suggests acknowledgement. Nonetheless, it is interesting that in this zone of concentrated state power and policing, the children remain somehow only minimally traceable, almost exterior to the machinery of surveillance, as a traitor within the king's court can be the least visible of perils.

Subjects, then, yes. Indeed the children are more subjects of and to the state here than they were in the village. But citizens? This is a more difficult question. While still within the village the children who became runaways were perhaps more citizens than they are in the railway space, where they become, as I offer here, subjects of a different sort of rule, discipline, and governance. From the perspective of the state, they should not and cannot be seen as citizens yet, not at their age anyway, even if they are people who will eventually become citizens through some process of passage. In its schizoid consciousness, they are simultaneously stains of shame, to be purged, and targets for a terrible sort of surrogacy, to be either coddled or corrected (see Cunningham 1991; S. Sen 2005; Foucault 2004; Ariès 1962).

The children certainly play with the notion of their potential (and unrealized) citizenship. In their own view they clearly have some claim and stake in the political idea called India, and they exercise its possibilities and push its limits. The stories of many permanent platform kids emerged during my time in Delhi, in which running away was presented as an effort to *obtain* social services and benefits not available in the village, or if not, to enhance social status, earnings, and opportunity until the point that parents would accept a transgressive child back into the home. This amounts to an insistent demand placed upon the state, if not for recognition then for rights, and it raises an interesting question: are railway children, to subvert Ong's (1999) formulation a little bit, flexible *non*citizens? Does situating one's self in railway space, in others' transi(en)t zones, afford a way for noncitizens to exercise flexibility and fluidity in their navigations of power (and in their noncitizenship)? To optimize the benefits they might reap and the possibilities they might realize, and to minimize constraints, depending on the circumstance and desire of the moment? To make the most out of invisibility and null status, to exploit all of its possibilities?

The petitions in Petu's case were a demand for a more proper recognition, under the law, of children in railway space, for acknowledgement of their presence as a starting point for intervention. Recognition has its perils, of course. It licenses a more sweeping criminal categorization of all such children and allows, in some readings, for a crackdown more thorough than any before. After all, if the law is to recognize children in such situations, urban improvers might say, then at least we will be newly able to cleanse the space, and at last with legal license. If the state recognizes such children, the recognition it bestows will certainly be deployed in the service of greater control, both of the children and of the aesthetics of space. Once recognized, they will hold no longer (or will hold even less fully) the right to exist. Railway children's lives thrive in the murky space between visibility and recognition. They circulate as they do in part because of the tension between their tacit acceptance and their administrative nonexistence, which is also, curiously, an existence suspended between city and country, between urban and agrarian worlds. But the best scenario, such planners and political aesthetes usually suggest, is to leave them unrecognized. To grant them legal recognition means acknowledging failures of the state and society better unmentioned, and mandating resources for new services and entitlements better allocated to outposts in Kashmir or Olympic bids or MMRCA combat jets. Let the kids scavenge the trash (say the architects of Delhi-the-Modern-Asian-City): giving them rights and making them citizens enshrines in law a repulsive landscape incommensurate with the one India is forcefully projecting and requires us to do something about them (which is in turn usually is addressed simply by deportation to avoid the new injunctions).

The notion of parentless children aboard trains as embodiments of illegality or criminality, and thus of something less than full citizens or members of society, or justifying their treatment as less than full rights-bearing citizens, is of course nothing new; solo children on the rails have long been associated with outlaws, brigands, and bandits. It is perhaps with the idea of such children as delinquents or criminals that society and the state (national and colonial alike) are most comfortable (see Radhakrishna 2001; C. Anderson 2004; S. Sen 2005; and Foucault 1995, 2004 on institutional productions of criminality). If such children are in any sense citizens, it is as criminals that their citizenship is rendered relevant: a citizenship of deviant externality to state norms that nonetheless forms an intimate relationship with the state. A citizenship defined by relationships to what is disallowed, by contrast to citizenships of propriety and patriotism, but nonetheless a relationship of fosterage and surrogacy by the state, with a telos that brings such subjects back into spaces of the state: corrective institutions of discipline.

The state has thus long interpreted and recognized children on trains through the lens of criminality. Almost all gangs of "railway thieves," wrote Rai Bahadur M. Pauparao Naidu in his *History of Railway Thieves, with Illustrations and Hints on Detection* in 1915, contained numbers of boys, and further, numerous of their adult members were either kidnapped or recruited as children (70). In the ranks of the Bhamptas, children kept guard and provided warning signals to adult members; sometimes children were employed as bluffs to allow the real robbers to escape. Among the Barwars, however, the boys would "usually do the actual thieving." A captive child was usually "pitied by the other passengers or bystanders owing to his tender age, and upon their interference [would be] let off with a slap or two" (33); the Koravars, he wrote, guessed that the "natural softness of the Indians will generally prevail, making them reluctant" to render to the authorities a "juvenile offender" (33). The boys were sworn to protect their superiors and never revealed "their places of abode or the names of [their] parents" (33).[6] Thus even a century ago solo and parentless children on trains were associated with criminality and with essentialized cultural traits.

So it is now, and so it was with Petu; the supreme court petition, in its review of relevant laws, makes it clear that the state and widely circulating popular ideologies still imagine railway children primarily as delinquents out of place and engaged primarily in a wrongdoing that violates rules for the proper and modern use of space. A criminal citizenship is what the law is likely to accord to railway children if it accords them anything at all.

Remand to Rehabilitation

Urban Institutions and Rural Children

How to Fix a Common Deviant:
Institutions, Runaways, Reform

David MacDougall's award-winning three-and-a-half-hour documentary *Gandhi's Children* (emerging poignantly on the heels of and in contrast to a prior documentary, *The Doon Chronicles*, about India's premier and most affluent boarding school) depicts in painful and painstaking detail the grueling, ruthlessly violent, emotionally oppressive daily lives of former street children, most of them runaways and all of them boys, in what appears to be a charitable home. We watch them arise and shower and weep and sleep, and beat on each other and get checked for scabies and compare rape stories, as sentiments are stirred and heartstrings tugged not by any narrative but by scenes themselves. It is almost too intimate, on some level, nearly voyeuristic, but it is hard to stop watching. The boys stare languorously and without affect from a half terrace, through an iron lattice, as giant pipes below emit a torrent of blackish sewage slurry (or something) into massive cavities in the earth with a force that would drown anyone in their path. The boys scream out at night in the halls, echoingly and achingly, for their parents to take them home. The scenes are engaging enough, and carry you along, until about halfway through, when something else begins to dawn on you: that what we are seeing looks like a shelter but is really a sort of prison.

Here the Delhi police, in a coldly brilliant sort of neoliberal innovation, have subcontracted with an organization whose original intent cannot but have been humanitarian (but nonetheless was founded by a policeman), to comply with a law that forbids street children from existing. How? By detaining them, thereby satisfying the demand that there *be* no street children, for now they are, as are slum dwellers in deportation cities fringing Delhi, like Bawana, *out of sight*. And thus, more and more starkly through the film, we see in the protective institution something that is darkly carceral, a subversion of the very notion of charity, from which children wish to escape rather than stay. There is, certainly, nothing new in this notion: if Victorian classics like *Oliver Twist* or *Jane Eyre* had a clear message on such spaces, it was that they are brutal, ruthless, inhumane spaces of abuse, not sanctuary. Furthermore, they drive home Foucault's observation of the twain history of the school and the prison. The chilling intimation here is that rural children, when alone in the city, must be held in a kind of detention camp to languish and stagnate or be reclaimed by the countryside, if the latter can find the child.

Unlikely. And they may well end up somewhere worse than this, in prison proper, and sometimes a place that looks perfectly friendly is a wolf in sheep's clothing. In all cases the children are singularly and curiously afraid of the institutions that claim to seek to protect them.

Invariably, among the first questions city runaways would ask me, while deciding whether or not to trust me, was whether or not I was with some kind of social service organization. They wanted to ascertain that I was not.

- What organization are you from?
- That NGO sonofabitch milked me for everything I had.
- The organizations? They just want to take all the kids' money and then use them to make more money.

These are some of the actual assertions made to me by certain of the child runaways who appeared to feel that the shape of their lives emerged from the evasion of organizations. The second one, the "milked me for everything I had," came from a child who had spent months if not years at a smaller, reputable organization, something more grassrootsy than most. I am making no assessment of the veracity of such claims; indeed, I think they need to be regarded with real caution. But they do reveal something of the degree to which these children want to avoid being tracked or located, enumerated or known. Why? Why should an assemblage, a genre of institution whose intent and foundation is at its core probably humanitarian, be seen to pose a threat?

Many possible answers arise, among them the one suggested by *Gandhi's Children:* that sometimes at their root such organizations' intent is *not* humanitarian, that they serve larger and less benign goals, goals of urban planners or states: "cleaning up" a city, say. I do not believe and neither do I wish to believe that is usually true. Another answer preserves the notion of their altruistic impulse to suggest that though such institutions mean well, many or most children who have arrived voluntarily "on the street" are there precisely to avoid what they perceive the organizations to wish to do, which is to confine and name them, saddle them with walls and unfreedom. In such a framing, the organizations are at the very least failing to author initiatives that correspond to children's goals, and at the very most they are close to jails, or homes with little to separate them from the reformatories that formed their antecedents.

One might ask whether it could really be true that such organizations may do the state's, the police's, or the city planner's bidding, and thus that they are in certain cases the carceral dressed as the charitable (and troublingly enough the true carceral, in the form of the state jails and reformatories, often dresses itself as charitable). But indeed it is true. In equal measure as nongovernmental organizations contract with the state to get kids off the street, two children in parallel situations might, by no coincidence, as well end up in proper state detention as at a charitable home like Prayas or Salaam Baalak, even if the term used is *rescue,* and not *detention.* Even more likely is that they will go *through* state protective services, via the Railway Childline and then the Child Welfare Committee (CWC), to a private agency that the CWC will have chosen for them. This is certainly not all sinister, and sometimes of course it helps the children in question, but it is often squarely in the service of the state, and what the state has in mind is not always and only the welfare of the child, and what the private organization to which the state gives the child has in mind is sometimes global donor money, just as the extraction of indigenous children from their families in South Dakota brought to the state federal money for each and every placement in foster care. In the eyes of some, furthermore, the subordination of private nonprofits to loose state definitions of child welfare embroils those nonprofits in the state's campaigns at the manipulation of space, of children in sticky cultural situations, and of slippery imperatives of modernization. Such parties claim that this implicates the organizations themselves in the violation of children's rights. In such a context the charity might be seen as nothing less than a subcontracting agency of the government that does the work of child detention and urban cleanup: a true neoliberal partnership.

Nonetheless, this is not true of all such organizations; many start from children's own perspectives, or at least from the notion of children's own perspectives. Many have produced amazing innovations, and thus remarkable assertions on children as citizens to be taken seriously, people with voice who can articulate more about their situations—and articulate it better—than outsider and adult interlocutors. Many far surpass what they promise.

This chapter focuses on institutional space, a domain with which nearly all runaways come into contact at some point. In the view of a broad range of institutions, "vagabond children" are defined as a "social problem," and in reductive ideologies this problem is often understood as a feature of the individual child—a feature, of course, to be "fixed." But how? The social or historical formation represented by "street children" is seen by a range of institutions of child management as a pathology of subjective abnormality to be corrected by discipline or other techniques. Indian solo children are themselves well aware that they are subject at any time to induction into spaces either of the state or of global ideologies of normalization, a prospect about which they are usually either guarded or outright scared. I wish to explore here the complex regimes of abnormalization and normalized aberrancy that govern the lives of these "anomalous" children. In particular, I differentiate between the "scientific" ideologies that are mobilized to motivate, legitimize, and justify the operation of institutions of child management. I am thinking here, for example, of the understandings of psychology and abnormality that govern state strategies of child confinement such as those at the "remand homes" of Sewa Kutir, and I trace the genealogies of global structures of "expertise" that encourage internationally funded Indian NGOs to seek to "correct" the "problems" of "street children" through an effort at middle-class socialization. If such children are just inculcated with the proper values, if they simply change their self-presentation, such institutions suggest, then their class status will be transformed and not only will they no longer be street children but the problem of street childhood will slowly disappear by virtue of new and wider social appreciation for disciplines of comportment and public space. Underlying this is a pathologization of social ills whose salve is identified in certain class-based truths of modernity.

Some such class-based possibilities, even beyond institutional programs of resocialization, are indeed organically part of what pulls children to the city itself to begin with, or of how its value is evaluated, after return. One recently returned child (Roshan) emerging from a stint at an NGO expressed some of the social-class magic he'd experienced in the city in terms of a perception of his place of origin as linguistically backward: In Delhi, he said, "people talk better. In Muzaffarpur they talk in Bihari. I like the language in Delhi. I don't like Bihari. My friends [here at home] are going to make me learn Bihari again." It is also of note that sometimes—again extra-institutionally, or more organically—a child's flight is narrated in terms of a moral, behavioral lapse associated with the city on the part both of parents and children whose "fix" is paired with being in the village: the now-grown former runaway Mohan evaluated his village's many runaway children in such terms. Those who leave, he said

> get in a bad company and start gambling. After they get in fights
> with family or others, they run away. The family doesn't worry
> too much. They get the money and they are happy with it. They
> feel it is better for the kid to leave rather than fight and gamble
> . . . For instance, my little brother ran away to Delhi. We ask
> him to come but he wants to stay in Delhi. He runs with bad
> company and he's gotten started with drugs. So he likes it there.

He will have to work hard and live properly at home. In the city,
he'll earn one day and then not work again until he runs out of
money. That's why he doesn't want to come back.

Thus running away is bound to a behavioral morality associated with
proper work, clean living, and good company, on the one hand, and with
home, on the other.

In asserting that a child's status as "vagabond" is often seen as an
individual problem to be fixed, by both rectifying the child and sending her
home, I am suggesting that organizations (and villagers themselves) quite
frequently frame the very *fact* of "street childhood"—and also of running
away—as an aberrancy, a psychosocial pathology stemming from some
fundamental, essential abnormality. Something, in other words, has to be
wrong with the kids. They're not right in the head, or even, they made the
wrong, uninformed decision because they are kids, itself a pathology, and
that needs to be fixed—either by reforming them, "rehabilitating" them
(that is, again, fixing them in a lasting way), or "repatriating" them (bringing them home, in other words, where they are seen to belong in a "normal" childhood). The way such abnormalities are constructed by nongovernmental organizations emerges in approaches to just how the abnormality
that manifests itself in existing on the street (or wanting to) should be re-
paired. Such notions are themselves, in turn, guided by globalized ideologies of childhood that have their roots in "the West"—in Euro-American
institutional norms that arrived in India by way of empire and associated
disciplines of capital, the entrenchment of schooling as norm, and missionary efforts (see Comaroff and Comaroff 1992); by way of more recent disciplines of global corporate capital and neoliberalism; and by way of global
flows of information and media through vehicles as diverse as advertise-
ments, films, and Internet pathways.

The notion, of course, of a "correct," "proper," or "normal" childhood, based on those norms of "the West," remains strong worldwide, in
part by virtue of colonialism but also by virtue of media machines under
conditions of capitalism that crank out the moralities embedded in adver-
tisements, films, and schools that make people around the world want to

embody and emulate them, that make people think they are right and much or all else is wrong. In front of me, on my desk, is a pamphlet produced by the Childhood Foundation of New York, a project of Queen Silvia of Sweden. It reads, in the slogan banner, "Giving children a childhood" (and elsewhere, "Will you help Her Majesty give children a childhood?"). In the United States, India, and other countries, institutions put a great deal of stock in spinning and constructing these particular ideologies of childhood—the idea that without a certain package of attributes and activities what children get is not in fact a childhood at all—and a great deal of effort into maintaining that model of childhood.

Schools, after-school programs, media outlets, religious institutions, policy bodies, and corporations all have a lot invested in prescribing elements of childhood, and in maintaining a certain normative vision of childhood as the correct one. These constitute a forceful ideology of what children *should* be. A childhood on the street, for example, one that involves distance from parents and home, one that involves consensual or nonconsensual sex, one that involves work, one that does not involve institutional affiliation—these do not, of course, comply, even in their least exploitive forms. The nonprofits at hand are at times hard at work to make sure they do. There are ways that such a thing can be good, and other ways it might not be.

The distribution of such a thing from the so-called West is not just a distribution in space but a distribution in time. Normative models have to arrive from somewhere, and some*time*, and so they arrive *in* the so-called West at (potentially) traceable moments as well. Philippe Ariès (1962) is famous for having pointed this out. This institutional construal—that the focus of correction in humanitarian approaches to street childhood is an individual child with whom something is wrong—begs mention of Foucault's (2004) construction of the "abnormal" in his eponymous 1974–75 lectures at the Collège de France, *Anormaux*. In these lectures, Foucault gives historical form to the meandering path by which individual children become a (or the) pointed focus of psychiatric power. He traces the transformations of the publicly accessible meanings of the "monster," the "onanist" (that is, the masturbating child), and most minimally (having run out of time in his semester) but nonetheless most important for this discussion, the "recalcitrant child"—the child who does not succeed fully in participating in society, the noncompliant "problem child." All of these contribute to a new and gradated psychiatric model of the child out of line who needs to be fixed with institutional power. All such children are seen to destabilize the proper social order; they are dangerous for the maintenance of order. Add to this his (1975) proposals on how discipline functions through technologies of bodily control and surveillance in institutional space from prisons to classrooms, and Foucault becomes quite potent for thinking on the institutional and carceral experience of Indian child runaways, and probably child runaways worldwide as well.

What is it exactly that is seen to be anomalous and abnormal about the runaway child? What is the thing to be fixed? Sometimes, as I mentioned, the child is seen to be psychologically "imbalanced," "not right in the head"—a problem that has what is considered an easy fix: resocialization, inculcation with right behavior, therapy. This in itself is of interest, as we cannot of course assume the cultural universality of therapy, of externalization, of talk as healing, which is based on a markedly particular Euro-American, and now globalized, model of mind and behavior very much operant in these institutional spaces. And yet, throughout India, ghost, jinn, and spirit possession (for example) might just as easily be engaged as an explanation for anomalous behavior, with exorcism, trance, or other spirit healing as its cure (see Freed and Freed 1990, 1993). I heard much indeed from the railway children, as I have indicated in chapter 5, on the distinctive potency of ghosts for them in city and village alike.

Sometimes, by contrast, it is the child's family—lapsed, irresponsible, unfit (a discourse tied to justifications for child removal in the United States as well, including from the South Dakota Indian reservations that I mentioned)—that is seen to be at fault and needing correction (see Scheper-Hughes 1992; and Goldstein 1998 for corresponding narratives of parental fault in light of poverty in Brazil). A participant in conversations in Rasalpur Block of Dumra, in Bihar's Sitamarhi, said, "The guardians don't really know where their kids are. So the kids are doing whatever they want." This could be an effort to underline what might simply be an alternate model of family, childhood, and supervision, but it sounds like blame. In the light of such imputation, rarely indeed is it the structures in which such parents might find themselves bound and under stress that are framed as being to blame.

But there is a deeper and more uncomfortable aberrancy constructed here. How—or *by what outward signs*—are the aberrancy and the abnormality that are framed as requiring management constructed? What is visibly aberrant, anomalous, or unnatural about the children, such that they cannot remain *as* they are and cannot stay *where* they are?

The intolerability of their presence in public space, for one thing, is a reflection of ideologies about both children and space. It is not just that such children cannot be experiencing what they are experiencing if they are properly to be considered "children" but that they should not be *seen* in public or consumer space. Their ragged physical appearance itself, furthermore, is rendered a blight on the urban landscape and gets framed more generally as a problem of personal comportment and "undisciplined" behaviors—the very ones, again, that can be fixed. These are themselves theorized in popular talk as a feature of a class identity equated with a flawed psychology. The disingenuousness, fault, and cheating that their orientation toward charity—and the world itself—is seen to represent emerge in such theorizations as a complex confluence of personal, individual abnormality and flaws inherent to the rural and poor castes and classes from which children are understood to come. This is how their class identity itself, irreconcilable as it is with visions of a clean modernity, a well-attired, market-oriented nation, or a capital- and commerce-driven city ready for global sporting events and multinational summits, is able to get framed as something that needs correction. In the face of this, the classes in question, not to mention the children understood to embody those classes in the city, have little chance to form any kind of solidarity that might allow them to share notions of an *us* misunderstood by a *them*, and yet and nonetheless they do, in certain ways, form such a thing.

And once the abnormality, the thing to be fixed (street childhood, aberrant urges to run away, the lapsed family), is named and given sufficient form, conjured, made manifest by its labeling, then the approaches at addressing it kick into gear. The kernels underlying these approaches usually correspond to the presuppositions underlying the initial assessment and reveal much about ideologies of childhood and social ill that in general underlie the institutions devoted to their amelioration. As one of the fundamental tenets of child aid is the redress of what are construed to be psychological problems and imbalances — abnormalities, again — the solution to such things is framed to be psychotherapy, western-style, and the provision of a stable environment, which of course may really enact a profound psychosocial transformation in a child's life. But beyond this, "solutions" for runaways range from improvements in clothing to computer-programming education, such as Firdaus pursued, or other occupational endeavors, from inculcation with very specific sexual rules and moralities to campaigns to get children in school and then college. Many of them revolve around norms of comportment: how to look, how to act in public.

It is not my intent to criticize or be cynical of most of what such organizations are doing — indeed many children, once they can save money, are able to go home, or change their lives in accordance with their own wishes, or escape disease or exploitation. What I am trying to do here, on the contrary, is not disparage, belittle, or devalue the (often life-saving) work being done, but to illuminate the unwitting and implicit ideological connections revealed in language between comportment and capital, between normalization and nation, between rehabilitation and culturally specific and widely circulating models of a "proper childhood," as interrogated in Ariès' *Centuries of Childhood,* as well as between the past's "reform" and the present's "rehabilitation." Some of these implicit connections are a feature of cooperation with state or police imperatives, or simply of social ideologies; others might have to do with the unending NGO need to garner donations. One Indian NGO thus speaks, for example, of helping children "enter" or "return to mainstream society" and engaging, vocationally, with "dream modification and correction" (in the case of unrealistic aspirations); a second speaks of "rehabilitation into the mainstream," and a third of "mainstreaming them into life"; yet another endeavors to teach principles of basic normative capital through "cooperative and financial management, fundamentals of management, accounting and banking principles and promoting entrepreneurship." One organization that emphasizes market-driven vocational training is also part of a government "Working Group on Reforming the Social Deviants and Caring for the Other Disadvantaged," and aspires to "promote national goals" including "national integration." A more detailed internal document from one NGO, consisting of records on individual case histories, refers to a child who "possesses the qualities of management and leadership." Many organizations indeed speak of socialization to nation and the disciplines of capital, of the "normal" and the "mainstream," and of self-comportment, class-transformation, and clean living as routes to success in these domains.

Why, one might ask, such class-oriented campaigns for such children, most of them rural by origin? Why endeavor to transform, in something that is almost proselytization, an individual's class persona in the face of an altogether different predicament of inequality, and rural-urban disparity, that is clearly in fact a structural and not individual ill? Perhaps beneath such efforts is an ideology of class transformation as personal transformation—away from abnormality and aberrancy. In other words, as part of a kind of prevalent meritocratic "Indian dream" (itself certainly present in Bollywood and legal reservations for the poor via "scheduling" alike), right class rectifies wrong comportment. But why? This starts to get at the question of what is deemed wrong with the sight of street children to begin with, to reveal something of why the image of the urchin—above all the urchin situated in class-marked space in which he does not belong—rubs society the wrong way. Class transformation as a solution to street childhood, furthermore, brings together institutional narratives of individual, psychiatric abnormality and larger tensions between the trope of the ragged child and aspirations of capital, modernity, and spatial aesthetics.

In other words, here is the question again of goals, of what kind of a "social problem" child runaways, "vagabonds," beggars, and other children out of place represent to those forces seeking to address something about their existence, or rather perhaps to redress something about their selves. What is it about their presence that mandates the frantic effort to fix something about that? Foremost, I suggest, it is their *visibility:* they are a blight upon an urban landscape that could, but for them, be the kind of

The new New Delhi: the DLF Emporio Mall, Vasant Kunj

modern space that its parent society aspires to, that can host international sporting events and erect gleaming shopping malls to match Dubai's. Much is to be gained from the image — whether in Delhi or Rio or Shanghai — that the city is capable and modern and ready for complete immersion in the skyscraper-inflected hypermodernity its competitors have achieved (Ghertner 2015). Delegates, above all, should not see rural runaways dressed in shredded tatters and sniffing solvent rags as they tour the town.

In considering the intent of such institutional functions, and the ultimate function of such efforts, in considering the question of why institutions would seek to rid the streets of runaways and then start by transforming their comportment, the matter of why it makes sense to labor to make a runaway a "normal" middle-class kid who goes to college and wears preppy clothes and does web design and volleyball and gets a job (and it is remarkable that a number of these organizations arrange marriages for their charges who in their solohood are seen as socially handicapped in this regard, for they have no kin to provide them a normative future), there emerges the question of whether and how charitable organizations' campaigns might unwittingly play into capital, production, power, and control. Do they—again unwittingly—help achieve the goal of cleaning the city of the unsightly, of the taint of poverty, and if so, why do they do it? To begin, I offer a remarkable example of the kind of discourse that sometimes gives shape to this narrow boundary between capital, on the one hand (and policing, to make a fine neoliberal consortium), and charity, on the other.

Runaways and the Beggarly Taint: On What Sort of a Problem Street Children Are

The social problem presented by street children, and thus by running away, that seems to need fixing is at least in part an aesthetic one, then. Such children should not be *visible* tainting the city. Delhi's reimaginings of urban space are the material realizations of a new notion of social order, itself rooted in certain longings for and projections of an ideal globalized modernity (see Harvey 1985; Harvey 1989; King and Kendall 2004); in such ascendant visions, space must be designed to accommodate the imperatives of consumption, and on all of its scales, incorporating the local consumer and the tourist, and also those who are in the market to consume Delhi itself: Olympic host city selection committees, Commonwealth Games planning commissions, global investors, planners of multinational meetings (that is, G7, GATT, WTO), diasporic elites attached to a certain conceptualization of what Delhi should be (see Ghertner 2015; Pablo Bose 2014 and 2015 (on Calcutta); Çinar and Bender 2007; King and Kendall 2004). And if this represents the actualization of a new notion of order, then (following Douglas 1966) it is subject to rupture and pollution, as order always is, subject to infringement, taint, and tarnish by the human embodiments of anomaly and aberrancy—beggars, urchins, vagrants, vagabonds, addicts, the maimed, and other such. If the new order is an ideology of affluence made manifest, then this tarnish is accomplished by the visible presence of poverty, especially embodied in "vagabond" (that is, aawaara or lawaaris) children.

Runaways beg. They don't just beg haphazardly or opportunistically, but regularly and systematically. They do this largely in places that are not only known for beggars but welcome them, even sacred places to which pilgrims travel to have the opportunity to give alms to beggars. Many or

Homeless youth and addicts in a culvert, Hanuman Mandir/Peti Market. Note that the young man at the center, in front of whom sits a small child, is holding a syringe.

perhaps even most child beggars are runaways. When Delhi, like most of India's and the world's poorer megacities, cyclically but perennially decides that it needs to be put on show and thus needs to look better, then beggars and thus child runaways are rendered a vulnerable target for expulsion and deportation. These are purges. It is my view that the effort to reform, re-move, rehabilitate, and repatriate street children cannot but be tied, even if loosely, to these imperatives of modernity, of getting such kids out of sight and off the street. The starting point for such a campaign, the thing that inspires it to begin with, is their presence *on* the street rather than, for ex-ample, larger historical or structural facts. They are only visible — nothing else — and that visibility is unhitched from anything else. It is in a histori-cal vacuum. When it is the case that beggars must be cleaned up it is often charitable organizations called upon to do it. And that itself is yoked to a more widely accessible discourse on the acceptability of begging in the coming age of a global capitalism in which India sees itself as a major par-ticipant (see Bardhan 1984; Sugata Bose 1990, 1994).

My analysis of this narrative of "beggary" is drawn from an unpub-lished but widely used text written by an academic team in the service of the Delhi government, and indeed molded to the latter's wishes, at its time.[1] But I use it as a token of a larger set of ideologies; this discourse is operant in municipal and national government but also outside institutional bounds among the elites who have invested themselves in certain ideas and imaginaries of Delhi, and it is furthermore not only a discourse, for I argue that it contributes to the actual shaping of urban space through its inflection of policy and its effect upon planners' decisions. At the broadest level, I suggest, this process represents the circulation of a set of ideas motivated by visions of the potential utopias that neoliberal globalization might herald, the possibility for morally good worlds that might be introduced by new forms of urban capitalism: I am interested in the conceptualization and interpretation of these ideas by their users as guides for authoring a new world. And such interpretation and conceptualization is not at all without consequence, and such authorship is not at all without its resultant texts.

But what *is* a beggar, anyway? It depends whom you ask.

In some of the older imaginings of the intersections of space and identity I am pointing to here, a beggar is just part of the social order; in some of the newer ones, in which, for example, Delhi might someday be an Olympic contender, beggars have no place in the presentable spectacular city, or in the city *as* spectacle (see Daniel Goldstein 2004). In such orderings, the beggar is aberrancy or social anomaly personified, as he represents an anomic state of things (following Durkheim and Simpson 1933) that must be denied, the failure of a society to attain the sleek and affluent modernity that is the object of desire (King and Kendall 2004; Ghertner 2015). To complicate things, there may now also be, furthermore, newly nuanced classifications of vagrancy: a police officer interviewed for this chapter, when asked how he would tell the difference between a faqir and a beggar, answered, "One can gauge it through their behavior by just seeing how they are communicating and behaving with others.[2] For instance, if a faqir is sitting in the park, he won't trouble anyone. He will sit in solitude, praying. Sometimes, he might ask someone for food if he doesn't have money, but he won't unnecessarily trouble anyone." In such formulations, we see the emergence of distinctions between a more benign, harmless, acceptable, and even pleasing mendicancy and a more insidious "beggary."

In 2007, I was given an unpublished "Report on the Survey of the Beggars in Delhi" authored by a professor of social work at the University of Delhi, Sneh Lata Tandon, along with her research team. The report (whose echoes of Kumarappa's [1945] *Our Beggar Problem* is striking) was funded by the Department of Social Welfare of the government of the National Capital Territory of Delhi (permission obtained from author for use in scholarship and publication). It encompassed data from 5,003 respondents. I include it here not because of the author's own views but because it reveals some of the discursive constructions surrounding the policy and practice of a type of "urban renewal" that has powerful implications for street-dwelling children.

The report presents beggary as a menace and a threat to public order, perhaps for the benefit of the expectations of Delhi's Municipal Commission that its authors imagined they might do well to meet. Tandon implores the audience to recognize that

> Begging is a problem from every point of view. It is first a problem from the point of view of the beggar himself. It implies for him a life of squalor and filth, of need and disease of ignorance and exploitation. . . . Begging is as much a problem for the rest of society. A large number of beggars means a non-utilization of available human resources as also a drag upon the existing resources of our society. Beggars are also a public health problem. They are often carriers of infection and disease. Begging primarily is indicative of the absolute failure on the part of an individual to function effectively and in a normal way within the society of which he is an integral part. (Tandon 2007: 2)

Much is of interest here. First, Tandon's suggestion that begging is the *cause* of "a life of squalor and filth" points to an understanding of the strategy as an errant or aberrant *choice* that has not been carefully enough considered. Second, the question of resource "non-utilization" underscores beggars as people out of place in the frame of the (imagined) welfare state. Third, much of what is presupposed here points to a concern over beggars as *bodies* of a certain type; these somatized subjects are the very vehicles of Douglas's "matter out of place": they are indeed social dirt. But they are more besides: in such a formulation, they are pathogen and pollutant—to society, to themselves, and by their own hand. The report states that they embody an "absolute failure" of a subject to "function effectively and in a normal way," and thus they represent, in Foucault's (2004) terminology, an embodied "abnormality," in part by virtue of a lapse in the proper "care of the self" (1988). Fourth, all of these assessments are assigned their value from within the disciplinary norms of a moralized capitalism: the failure, for instance, of "function" is a failure of body and self to comply with every-day behaviors in a market-inflected space. There may be buyers and sellers and viewers, but in such a frame beggars "drain."

It is this perceived failure to fit in regnant relations of production and consumption that comes out, for example, in the report's interest in dis-crediting the many "able-bodied" from among a "beggar population who should normally be working for their livelihood." This moralized capitalism emerges in descriptions of the ways that society is understood to be duped and deceived by the unitary mass of insidious and cunning indigents: "The beggars," writes Tandon, "have their own ingenious way of appealing for alms. . . . Begging is an activity, which needs to be discouraged as it has many evils." Tandon and her team in their recommendations for policy fur-ther underscore the manifold "evils of almsgiving," emphasizing that the "public should be informed of the evils of almsgiving. That giving of alms promotes parasites in society and de-motivates people from doing hard work." Thus again the aberrancy of begging is narrated as an individual's failure of compliance to duties of industry and industriousness within re-lations of production.

Neighborhood residents in Delhi's Nizamuddin, a historical center for begging and a critical meeting point for runaways, have similarly harsh views on beggars. A conversation with a tea-stall owner and an ear cleaner unfolded as follows:

TEA-STALL OWNER: Even if you build them a house and give them, they will still beg. Even if you keep them in a palace, they will still beg.

EAR CLEANER: Beggars in this area are from top families, well-off families. Even if they have a crore (ten million) rupees, they will still beg. People as far from Patel Nagar come here to beg despite having good business. They are in the habit of begging.

TEA-STALL OWNER: Even if you get them a flat somewhere, these people will continue to beg. They will come out of the flat and beg. That is why nobody cares for these people. Wherever they go, whether Daryaganj, Jama Masjid or elsewhere, they might change the place but they will still beg. People who do not get habituated to doing work or business, and get money for free and prefer to beg only. He will never work and will beg throughout his life.

The view of begging, then, as an "absolute failure" of industriousness, labor, and merit—as a personal rather than a structural failure—is not an element of institutional ideology alone but also one of popular discourses. A day laborer working as a "cleaner" commented that he makes a living by working hard and earning while the beggars earn without working. Indeed, so widespread is this narrative that even several beggars and other itinerant subjects commented on some other beggars' undeserving status.

Just as "able-bodied" beggars are seen as undeserving and truant from their proper place in labor, so too are children seeking alms seen as out of the places normative views of life cycle and (middle-class or welfare-state-poor) childhood accord them. In general, this is the discourse that frames public discussions of street children, but in the discussion of begging the emphasis on their responsibility for their status as beggars is further and more forcefully emphasized. "Most of the child beggars," the report explains, "were of the age, in which they would have enjoyed their childhood and school life, but instead of that what were they doing? *They are spoiling their childhood.*" Thus groupings of children such as those who gather in the central square of Hazrat Nizamuddin just west of the Bangla Wali Masjid are spun as miscreant failures of their own making, just as later they will be dropouts from their proper place in labor relations. For this they must be punished, as a complicated array of laws stipulates, or put into one of several charitable programs or shelters ("to give children a childhood": see Hecht 1998; Glauser 1997; or Scheper-Hughes and Hoffman 1998) whose implicit objective is essentially to socialize them to school-going, extracurricular activity-pursuing, well-dressed, middle-class childhoods.

According to the 2007 report, laws seeking to imprison beggars are framed as insufficient for the "reform and rehabilitation" of the beggars. Among the recommendations to address these various lacunae are policies for "rounding up the beggars." Rounding up and deportation is widely seen as the most effective strategy for cleansing space of undesirable subjects in advance of cleansing the space of its undesirable material infrastructure and aesthetic (following Ghertner 2015).[3] In Nizamuddin, explained a police officer:

> for cleansing, the Municipal Corporation of Delhi [MCD] brings its van according to their schedule every few days to catch beggars. But beggars also have become smart. They know they will be picked up by these vans and taken away. There is a fear of where they might or will be taken. They get food here conveniently and don't know whether they will get it outside or not. The Delhi Urban Shelter Improvement Board has constructed very many *rain basere* [shelter homes]. They have made shelters for homeless in quite a few places. We also encourage them not to sleep on the roadsides. We tell them whatever you do during the day, spend your night in the rain basera. We encourage them morally. We cannot beat them up. Take away their *rozi roti* [daily bread] instead. . . . Even MCD does not do anything coercively. Rather it takes them to the right place: provided by the government.

Locals have their own view on this, however. As a tea-stall owner in Nizamuddin said, "Whoever tries to remove poor people will himself be finished. With my tea stall on the footpath, my life with my ten children goes on. Whoever removes us, we will curse those people: the Aga Khan, or some minister, MP, or whoever." A beggar woman commented that "beggars from the area are removed when a politician or a film personality or any famous rich person visits the *dargah*. So a word comes from the dargah to leave the place for an hour or so. Sometimes a policeman comes and tells us to leave for an hour or two. Even if a high-level policeman is visiting, we are told to leave for a few hours." Major global events occasion the expulsion of beggars, homeless migrants, "street children," and as here, even unwanted residents to containment zones like Bawana, an area of concrete tenements in the far north of Delhi's periphery where deported slum dwellers and homeless people are periodically forcefully relocated.

The trajectory of slum clearances at the Yamuna Pushta in advance of the Commonwealth Games illustrates this well. The first demolition order was issued in 2002, for a "Games Village" and a "green belt." The very first resettlements occurred in 1967. But in 2004 the Slum and Jhuggi Jhonpari Department of the Delhi Municipal Authority had issued an eviction order. Many of them were resettled to Bawana. During the three months of the premonsoon season of 2004, twenty-seven people died in Bawana alone from causes related to public health and sanitation. By 2007 an eight-lane highway had been constructed, four hundred thousand residents had been relocated, and eighteen thousand slum clusters demolished (see Menon-Sen and Bhan 2008; Verma 2002). By the time I walked down to the river-bank that year (2007) at the Yamuna Pushta, the masses of homeless children and adults had returned, but the slums were still gone. This was not far from Nabil's 1852 Calcutta Gate. Likewise in 2011, though by this time settlers in straw huts were cultivating the shoals in the middle of the river. Grassroots advocates whom I spoke to described the relocation villages in Bawana and elsewhere as vast and desolate zones of containment and detention.

The report pays more than lip service to this imperative for which, it now becomes clear, it was indeed written. Acceptance of beggary is rendered opposition to progress: "The menace of beggary in the National Capital Territory of Delhi has been of concern to all especially in view of the forthcoming Common Wealth Games in 2010. Hence the Department of Social Welfare, Govt. of NCT of Delhi decided to get a survey of beggars conducted with a view to know the exact profile of the beggars so that planning for their rehabilitation could be formulated" (Tandon 2007: 15). And indeed, the report's recommendations were implemented, or at the very least they were mobilized in the justification of that implementation, a report to legitimize the expulsions.

By 2008, in preparation for the Commonwealth Games, and potentially in response to this report or out of the same directive that issued it, orders for the relocation of Delhi's beggars had in fact been issued. Newspapers reported that "the Delhi government Wednesday informed the Delhi High Court that it was ready to hold mobile courts to make the capital beggar-free by the time of the Commonwealth Games to be held here in 2010" (*Tribune India,* May 8, 2008). Policymakers suggested that the "initiation of measures to send beggars to their native places from where they have migrated to the capital will be a another good option to stop the menace of begging in the capital" (*Tribune India,* February 19, 2008), echoing nearly precisely the report's language. The aftermath of this, according to a Guardian (UK) report (November 8, 2009), was the implementation of the spatial cleansing as intended:

> After they were locked up in beggars' prison behind the high, barbed-wire-topped walls of the Nirmal Chhaya complex, next door to Delhi's Tihar jail, 50-year-old Ratnabai Kale twice tried to hang herself with her own sari. As India's capital stumbles towards the starting line for next year's Commonwealth Games, draconian orders have gone out to clear the streets of beggars. Teams of police, backed by mobile courtrooms, are roaming the city, dispensing summary justice to those whose faces don't fit. The rationale for the purge is simple: the image of an outstretched hand does not sit easily with that of the "Incredible India" that the authorities wish to project. "Before the 2010 Commonwealth Games, we want to finish the problem of beggary from Delhi," the city's social welfare minister, Mangat Ram Singhal, announced at the launch of the initiative. (Chamberlain 2009)

Again, the report's language echoed in policy itself. In sum, then, the messages here seem to be: the vagabond, beggar, or urchin is culpable; the beggar is exhibiting a pathological behavior in need of discipline; the beggar disrupts the potential for harmonious spaces of consumption; and the beggar's right place is in industrial labor. And thus in efforts to produce a modern city, a runaway at an intersection becomes a synecdoche and an emblem of something much larger, and something pollutive. And thus the child *being* on the street, not the reasons he got there to begin with, becomes the policy imperative. And who better—and less costly—to help clean things up than charities?

Deep Structure: A Note on Disparities in the Rural-Urban Institutional Divide

What the Foucauldian paradigm fails to account for here, of course (as elsewhere), is the structural architecture of society that puts children in the city to begin with. For that we must turn to Marx and others concerned with structures. In a curious correspondence to this material realization or reflection of theory, organizations seeking to change *structures* associated with children's trajectories to the city, rather than mounting fixes that address their appearance and existence as "street children," are vanishingly rare. What kinds of structures could be changed, and are such changes something that can or should happen? Even if those changes happen on the small scale, do they really do anything to prevent a child from leaving for a "street" where he will be raped, sold, sickened, decimated by inhalants, or killed? What if the prevention of the child's departure puts him in a place equally miserable to that, or nearly so, or worse? Then should structures be changed such that children do not leave home? Only if they somehow ameliorate the misery (slavery, abuse, daily subjection to death) itself, of course, such that one place is rendered less miserable than the other.

Nonetheless, institutions *do* exist out there in India's vast expanses that incidentally (if not specifically) seek to change some of the same rural structures that make it intolerable for some children to stay home, though seldom is the stated goal tied explicitly to running away. Notable, however, in its child-centered rural effort is the Concerned for Working Children (CWC) in Karnataka, thrice nominated for a Nobel Peace Prize, an organization that pioneered children's formation of labor unions—for children—and that facilitated children's formation of their own structures of rural governance, Makkala Panchayat (children's councils), Makkala Grama Sabha (children's village meetings) and Makkala Ward Sabhas (children's ward meetings), all of which also helped increase *adult* participation in governance. CWC is also at the cutting edge in its success in mobilizing children themselves to do research—about children like them. Likewise, Nobel Peace Laureate Kailash Satyarthi's Bachpan Bachao Andolan (BBA) works extensively and quite successfully in the countryside and legislatively to engage and transform conditions and norms that lead both to child slavery and to children's entry into labor. This of course has substantial bearing on runaways' situations and prospects as well, in more than one sense. At BBA's hands, the better part of a hundred thousand child workers have been restituted to their villages or placed in safer conditions elsewhere. BBA has produced substantial change, including in basic laws, though it probably has not changed the fundamental global and national structures—themselves emerging out of what Carolyn Nordstrom (2004) calls "shadow" markets—that generate demand for the purchase, sale, and traffic of children. As I have described elsewhere, these are some of the same forces that propel children from their villages to cities, either because they are fleeing that or because the stresses that would predispose a family to sell their child are the same that make staying intolerable.

Elements of foundational modern Indian agitation from Ambedkar to Gandhi might be found in the discourses, ideals, and roots of such organizations. Some of these efforts, furthermore, are surely inspired by or emerge out of the same radical ideals as the Grameen Bank, in its original form (see Ananya Roy 2010 on "poverty capital," "millennial development," and the economy of expertise on the management of the poor); as grassroots movements like the Rural Litigation Enlightenment Kendra (in the Indian Himalaya) or the Narmada Bachao Andolan (in Central India), both focused on the nexus of people and environment; as Vandana Shiva's Navdanya, which focuses on counteracting rural debt and the iniquities tied to Green Revolution high-yield varieties and on reintroducing native "seed sovereignty"; and as Brazil's Landless Worker's Movement (MST). MST, a grassroots organization fundamentally engaged in land reform for the rural poor, informed by both liberation theology and Marxism, follows a "local council" structure, with its small, settlement-level Nucleo de Base organizational bodies embedded in wider forms up to the national level. Such councils as a component of development efforts are quite common in South Asia, including in the Aga Khan Foundation's rural campaigns (see Steinberg 2011), but also, as mentioned, in children's organizations.

Among these is Bal Sakha, an organization working in critical run-away-dense areas of Bihar; I was struck, when I met its founder, Sanat Sinha, at the attempt to mobilize indigenous structures—including literal structures used for gathering places based on traditional architecture from village centers—to encourage children to form their own councils (in this case called Bal Suraksha Manch). In urban areas including Delhi, the Butterflies organization runs the so-called Bal Sabha children's councils in bigger multineighborhood formats, wherein child representatives from different neighborhoods come together to discuss shared concerns, and in smaller formats, where children from a single neighborhood discuss problems they face and possible solutions. I had the chance to observe a large Bal Sabha in Delhi's Nizamuddin neighborhood: remarkable. The children brought to the floor an exceptionally cogent set of pressing and critical concerns and impressively conveyed the sense that they felt themselves in charge of, well, something. The organization and process were astonishing.

What is the ur-form that underlies the interest in the creation of village councils, and children's councils more specifically? What model do they work from? Is there a prototype that is being replicated? Certainly it emerges in part out of a certain global dissemination of sometimes-triumphalist ideologies of modernity, out of norms of democratic participation, civil society, inclusion, pluralism, and liberal humanism (that are nonetheless themselves associated with free markets, with capital, with contribution to labor): everybody contributes, everyone's voice gets heard, and thus everyone has a certain type of citizenship in a certain type of improvised, conjured space. It is not a state, not a settlement, even, usually, but it is a space of political decision-making, of sorts. Do such child-centered (or village-centered collective participation structures) constitute a discipline, a form of power, too? Yes, certainly — they provide a space for the exercise of power by some over others, a context for surveillance, and a forcefully exerted technique of control in the Foucauldian sense (with the consequent right ways to sit, speak, participate, raise your hand, and lead). But such a council also might provide a space for the production of counternarratives, for certain forms of defiance, for the alleviation of suffering or the rendering of secrets or shared social realities. Such critiques need not suggest that these forms cannot also be good, even when historically speaking their origins and intentions are complex.

Do such organizing efforts in the countryside ultimately do something to change things, or, as a kind of romantic and sweetly appealing idea, do they just look and sound nice? I think they are more than that, more than pleasing photo ops for kids' democracy. When everyone is dying of pesticide poisoning, say, or floods, or leishmaniasis, such efforts aren't really guided by triumphalism but, instead, by urgency. Might the collective naming of a problem such that it might be expunged make a child's village life a little better? Might the very act of allowing a child to do such a thing in the face of its rarity not have its own psychosocial value? From the Rawlsian perspective on justice, or in the elaboration and critique thereof by Sandel (2009), if someone's own definition of goals is being valued, if choice is strengthened, then maybe something is being done from the perspective of someone who wants it done. A message to render from this is that it is a little too easy to be skeptical, perhaps, and on what counts and what works we need to listen to the children themselves.

Nonetheless, such things are already part of the landscape from which children continue to choose to leave and do not seem to suffice to prevent their departure. One is compelled to ask: is there anything that would prevent their departures? Should there be? Perhaps not, if framed like that. But there are things that might transform the miseries that spark in them the desire to depart. Could those things be changed, structurally and widely? That would be a grand and difficult task, and the danger therein would be the risk of always ending up with the suggestion that in order to do that it is culture itself that must be changed. Such a suggestion has been made before, and it has a troubled past.

The God of Urchins: Sacred Sites as Humanitarian Counterinstitutions

Lacking the carcerality, the corrective feature, though sometimes exposing children to certain campaigns at inculcation, conversion, or recruitment is a whole array of something else, something remarkable, that doesn't appear to be the same thing but that in fact for the *children* fits into the same map, the same calculus, of sites from which they might receive services. This something else has an altogether different cultural genealogy, a history much deeper and thicker and, indeed, more autochthonous and more local, fundamentally speaking. Interestingly, or tellingly, perhaps, this something else interacts but little with the sort of structures that the world of children's institutions is made of.

These sites of profound child refuge are sacred spaces of one branch or another of faith. As everywhere, I ask: must they all be seen as the same thing, or is some categorical unity imposed on them? In the eyes of the children, in my view, they do appear as a unitary sort of thing, a nodal web in which a person circulates from one thing to the next by virtue of this sameness. This is an altogether different sort of humanitarian space at which child runaways converge, one that is not at all an institution in the modern sense but nonetheless provides services. I went around to many of these and eventually found in them the best place to meet runaways. But these sites too are considered taint and stain because they draw the poor and thus require cleansing.

Such places in fact gain something from this as well, even if they remain external to the world of *global* alms, the ever-present imperative to scrounge funds from international donors that pervades the NGO world — but what they gain is largely in spiritual capital. In all or most cases, such humanitarian sacred spaces' existence emerges out of injunctions on providing alms to the poor. In such spaces a sort of exchange thus takes place: people come to find poor people to whom alms might be given. This is auspicious and beneficial for the donor. Simultaneously, the poor—among them many runaways—flock there to receive such alms, and their flocking is itself not without substantial knowledge of the spiritual benefits to be gained from being in these places.

Perhaps the busiest places for such a dynamic in Delhi are the Sikh *gurudwaras,* with their famous *langar* charitable kitchens, a concept with roots in Chishti (Muslim, non-Sikh) Sufism. The Sikh langar fills, at mealtimes, with the needy, among them many runaways; these visitors receive a free meal and must in return provide some service—cleaning up, helping serve others—at the temple. The rhythm of the Sikh langar, indeed, guides many children's movements through the day, such that runaways I knew at Nizamuddin shrine, at Nizamuddin Station, and at New Delhi Railway Station would up and leave their normal haunts in the middle of the day, saying, "Hey, let's go to the gurudwara." In Delhi this means, above all, Bangla Sahib at Connaught Place and Sis Ganj at Chandni Chowk, in Old Delhi.

In some cases the accommodation of the poor at a sacred space draws a semipermanent population. At the Hindu temples of Prachin Hanuman Mandir (the Old Delhi one near the Yamuna and Peti Bazaars, not the Connaught Place Hanuman Mandir) and Kali Mandir, also called Kalkaji, the ability to give alms to the poor is an essential part of the pilgrim's experience. Hanuman Mandir, accordingly, has a large group of resident runaways. This is where I found Nabil, Deepak from Buxar, and many others. Some of these children find a patron shopkeeper for whom they do jobs in exchange for a bedroll and place to sleep at night. Here at Hanuman Mandir is also a set of makeshift movie theaters—teahouse huts with televisions—where many of the children gather. On holy days, Tuesdays and Saturdays, the temples are flooded with beggars who flock in from farther afield to line up to receive charity from the templegoers, and there are soup kitchens in the complex.

The charitable infrastructure is even better elaborated at Kali Mandir, which has a number of soup kitchens but more notably a very stable set of permanent runaway residents. Here a field serves as a community living space for a large number of children. One child, Balram, would admit that he ran away but would go no further, looking stone-faced and sad at any

Beggars line up for alms, Prachin Hanuman Mandir near the Yamuna River

further questions. His dogged silence was remarkable. Another, in 2015, was uncannily similar in his stonefacedness; he had just arrived from Bihar, having only recently run away, and seemed a little shell-shocked at his new community. But, somehow, he had heard that this was the place for him to come live for a while and be provided for. Another child, Prashant, lived here with his mother and begged by day but had himself run away *from* here several times. While I was standing there in the summer of 2015, the temple children were working to accommodate a disabled child of perhaps sixteen years, unable to speak, who had been left there by his family with a note in his hand that read "please help me."

I was surprised at the way in which the children had all rallied around him. Most remarkable, however, was the noodle and dumpling house run by a gnarled-handed man from the Himalayas that provided shelter and employment to dozens of boys who slept on its floor by night and worked its cauldrons by day. Two of these were orphan brothers from the south, from Madras (which they pointedly did not call Chennai). Several others were runaways from Bengal. I was astonished to find that these boys that

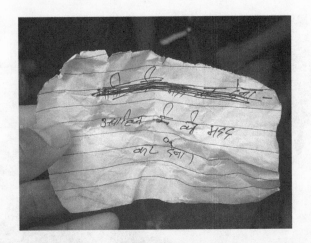

Kalkaji, "Help me." The boy to whom the note refers is at the center of the gathered children. The boy immediately at his feet is a new arrival, a runaway from Bihar.

I had known in 2011 were still there, now men, in 2015, working the same space—a tremendous continuity. Even in 2011 I met shopkeepers who claimed they'd run away as children and settled in the temple complex, still there after all these years, even at that point.

The grounds of the Jama Masjid, the congregational mosque of Delhi (and India's largest), with the vast maidan and the tent cities lining the walkways and empty fountains, are perhaps the largest refuge for runaway children in India's capital by number. Here is where I met Firdaus, Akhil, and Shamsuddin, whom I discuss elsewhere, and where Kaalu died. But conceptually speaking the true champion of informal sacred humanitarianism for runaways, and the one evincing the best-elaborated theory of aid to the poor, is the Sufi shrine (in general), the space in which the langar originates, rooted in movements at the base of whose mandates is care for the destitute and forsaken of any religion. Nowhere in Delhi is busier with beggars than the labyrinthine neighborhood surrounding the *dargah* (tomb) of Hazrat Nizamuddin Auliya, a fourteenth-century Sufi saint (b. 1238, d. 1325) of the Chishtiyya order. Hazrat Nizamuddin, as the area is called, has been built around the idea of almsgiving, around which the neighborhood's circulation has thrived some seven centuries (see Dhaul and Kamath 2006); its history is inseparable from the economy of begging; its interactive fabric revolves around an encounter between pilgrims and the poor.

But more specifically it is a begging prescribed by *Islamicate* visions of who begs, how to beg, who gives, and where to give, imbricated in the neighborhood's spatiality. Such a vision comes not only from the Muslim emphasis on zakat in its directly charitable forms but also from the historical Nizamuddin's insistence on the primacy of provisions for the poor of all religions (see Lawrence 1991). This insistence has antecedents in the teachings of the founder of the Chishti *silsilah* (literally "chain," a Sufi order) in South Asia, Khwaja Muinuddin Chishti (1141–1236), and in the teachings also of Nizamuddin's own teacher, Shaykh Fariduddin Ganj-i-Shakar (1173–1266). The shrines of these saints (the former in Ajmer) are similarly characterized by very active populations of beggars and runaways, and familiar features including the langar (the charitable kitchen) and the *degh* (the consecrated cauldron). Both the saint and the shrine at Ajmer, in

Rajasthan, are called *Gharib Nawaz*, the comforter (or benefactor) of the poor. Upon visiting this latter tomb at Ajmer, I was astonished to learn that many children run away directly from home to the shrine, as the British cantonment that was here brought a full-service railway line that can convey children in flight right to the saint's place of rest. Indeed, I met at Ajmer several children who'd come from West Bengal directly by rail.

Indeed, the Chishti order more than any other has imprinted itself on active Islamicate space in Delhi, or on space *seen* as Islamic. Thus it is a practice of Islam contextually tied in to the northern subcontinent, and to Delhi itself, around which discourses of begging in these spaces have evolved.[4] Perhaps most essential is the Chishti emphasis on the Sufi devotee's voluntary assumption of a life of poverty, and a related commitment to society's poorest classes, Muslim and otherwise (see Ernst and Lawrence 2002: 3).[5] Ernst and Lawrence quote a fifteenth-century Chishti saint, Simnani, to characterize this orientation: "They aim at poverty and denial," wrote Simnani, "and they keep company with the poor and beggars, giving them food" (2002: 6). Likewise today: Chishti tombs assume the saints' practices, providing for the dispossessed and desperate.

Because this is such a central node in runaways' circulation not only in Delhi but even on the scale of India itself, it is worth dwelling on what sort of space it is. A pilgrim might choose to enter today's Nizamuddin *basti* (a term denoting an old-style neighborhood, and also sometimes a slum) from the east side, at the entrance on Mathura Road adjacent to the Barakhamba Road police station, among many other entrances. Past the police station is a field often full of sleeping beggars, but at my last visit it was being "beautified" and paved, a sure harbinger of the impending expulsion and continuing nonadmission of such people. Neighborhood shopkeepers and other inhabitants in the informal economy most often ascribe a criminal identity to the itinerants of the neighborhood, labeling them primarily as pickpockets and addicts (and thus undeserving). One of the khadim caretakers echoes this, saying of the local poor (or those so perceived), "These people are not those kind of people who are needy and need alms. These are not the needy sorts. They are involved in criminal activities and drugs, snatch people's cell phones and pick pockets." Among those undeserving are surely any children who may have chosen to leave home.

Slum dwellers, Nizamuddin urban forest

Just to the north, at the large *chowk* (an intersection or crossroads), many runaways beg (along with others) among and between the cars when lights are red. To the east, now at the visitor's back, is a scrubby forested area housing hundreds of homeless migrants, most of them beggars, some runaways, some street-working children, some in tents. One resident expresses that once "this was all forest," bemoaning its pollution by undesirables. In that direction is also the affluent neighborhood of Nizamuddin East and, eventually, the Hazrat Nizamuddin Railway Station, home to the dozens of runaways I've mentioned who subsist largely on a trade in plastic bottles scavenged from trains. Still beyond that is the Sarai Kale Khan slum, and eventually the Yamuna River. The presence nearby of Humayun's Tomb is deeply important, as the emperor's grave was built expressly to be close to the dargah, and the dargah's restoration follows on the restoration of Humayun's Tomb.[6]

Back at the described entrance to the *mohalla* (neighborhood), at either side of that gate, wait a number of beggars alongside tongas and rickshaws and pilgrims awaiting rides. The same is true of the wide passage conducting pilgrims in the direction of the tomb. On either side of it are numerous beggars, mostly children, women, and mutilated men. On the right side of this passage is the Bangla Wali Masjid, the center of India's Tablighi Jama'at, a Hadith-based witnessing or storytelling movement of anticolonial North Indian origin that is now perhaps the largest Muslim network in the world (see Metcalf 1993; Gaborieau 1999; Masud 2000).

Beyond the Tablighis' Bangla Wali Mosque one enters a small square at the convergence of a tangle of paths, a kind of pedestrian crossroads where motorcycles are also parked. Here are a number of "charitable soup kitchens" that sell to passers-by cards that can be given to poor children and to others among the indigent and destitute, to be exchanged for food at the kitchens (and though the kitchens really do sometimes feed the children, the trade in these cards is in part a scam, as once the cards are sold and the money collected and the visitor disappears the cards are usually redistributed to the restaurants and passed back out for sale without the distribution of a meal in every instance, to create merely the appearance that a meal has been donated).[7] By virtue of the presence of this subeconomy revolving around the visibility of poor children, and of the flow of people through this open space, this is the gathering point for many of Nizamuddin's solo children (largely runaways, with some orphans and also some migrant children with a parent nearby). These children spend much of their day at this crossroads, moving sometimes to the shrine, sometimes to the small video game arcade down the alley to the left; they are largely from Bihar and Bengal (though I have met children here from as far away as Manipur and Tamil Nadu) and mostly but not exclusively boys.

And they too figure into what is perceived to be preventing the ascendant ideal of a glitzy and global Delhi; thus they too are among the subjects targeted for removal.

One might proceed down another alley, tighter now, toward the shrine, among pilgrims from around the world. Nonpilgrim tourists from everywhere will also be among those processing toward the tomb. In this section pilgrims buy garlands to decorate the saint's tomb, sweets or incense to place there, hats to cover their heads, mementos. All of these are promoted vigorously by hawkers in stalls lining the way. The visitors will be asked to remove their shoes at just *this* stall, to purchase flowers to show

The Nizamuddin video arcade near the Beggars' Crossroads

the requisite devotion. By now the visitor will likely have given alms to a beggar: it is an integral component of the experience, and necessary to ensure the dispensation of the blessings being sought at the shrine. Many or most of these features are found in sites of pilgrimage across South Asia and are not unique to this one. In this area one passes the important tomb of the seminal Sufi-influenced late Mughal- and early British-era poet Mirza Ghalib.[8] The tomb remains an important site in the context of the nearby Ghalib Academy and was recently renovated by the Aga Khan Foundation.

Finally, at the paths' narrowest point, the walker reaches the tomb itself and removes her shoes, if she has not yet done so. Here sit the shoe collectors like Shahzad, a runaway from Assam, and more small stalls selling pamphlets and music and flowers. Through a low marble gate entry is granted to the shrine, and the modality of space changes utterly.[9] Surrounding the edge of the large white marble enclosure of the shrine are centuries-old jali lattices, also hewn of marble, obscuring graves of the saint's relations or disciples, or of lesser saints. Also arrayed around here are small stalls serving as contact points or offices for various descendants of the saint who are shrine functionaries and involved in the management of the site. There is a dispensary and the free kitchen, the langar, to serve the poor, again reflecting that Chishti emphasis on compassionate engagement with the suffering. Near the entry is the shrine of Amir Khusrau, one of Nizamuddin's disciples and a founder of *qawwali* devotional music, a self-sufficient subsite of pilgrimage for many, populated in part by Khusrau's descendants, into which one is urged to enter before visiting the rest of the shrine.

The tomb itself, where Hazrat Nizamuddin Auliya is buried, is the center of human circulation and orbit in the shrine enclosure, and the whole basti is an effervescent site bounded by shining, silvery jali lattice screens; surrounding the domed tomb is a verandah-like platform, punctuated by pillars. The tomb stands in the middle of the enclosure, surrounded by a courtyard that is the site of intensive ritual activity, including qawwali sessions, prayer, contemplation, and begging. The whole configuration echoes Mughal courts and gardens alike, including the *charbagh*.[10] At one edge of the courtyard is a Mughal-built mosque in which *salat* prayers happen at the assigned times. Homeless residents, including children, are welcome to stay the night in a chamber on the left just before the entry to the mosque. Women may not enter the central part of the shrine, the actual tomb. Men and boys enter, dispense garlands upon the shroud, kiss the shroud, kiss the stone beneath it, place their heads beneath it, circumambulate it, and perform *du'a* (prayer in its beseeching form) within it. Some will be weeping as they circumambulate. The inner dome is adorned with inlaid mirrors. Men may take photos within the tomb enclosure itself; I have seen them deliver the image to the women who are awaiting them outside the shrine, whereupon the women kiss the *image* on the mobile phone as a virtual way of accessing the tomb.[11] Women have another channel by which to access the tomb: a jali on one side of the tomb into which they can peer and before which they sit in reverence, facing the direction of Mecca. On the verandah surrounding the tomb sit those men who are chief among the saint's descendants, the ones who purport to be able to demonstrate the most direct line of inheritance; there they vigorously exhort visitors to make a contribution, and in return the visitor may receive in any country a lifetime of printed newsletters, which I did for over a decade.[12]

A boy resting in the Vagabond's Passage, Nizamuddin shrine, 2015

The grounds of the enclosure are full of working children who walk around to provide more flowers to sprinkle on the grave, while beggars walk the crowds (most dense on Thursday nights, festival nights, and the saint's *'urs,* or death anniversary) and ask for alms, particularly in the stairway passage to the right of Amir Khusrau's tomb, on the way down to Nizamuddin's. Some of the resident poor who perform these tasks also have the job of fanning the devotees on hot evenings. If the visitor passes just out of the enclosure in the same direction he would have faced when entering the tomb, he enters a long arcade (not the same as the other enclosure mentioned above in which the homeless are invited to sleep) running perpendicular to this direction, adjacent to another mosque, wherein many homeless people, including the beggar children, sleep at night, and they are welcomed. Nizamuddin's injunctions (see Lawrence 1991) decree total acceptance of the presence of beggars in his abode; in between prayer times, the working children and beggars rest in the marble enclosure, play, sleep, eat. They are tolerated fully and without condition; their company is even invited.

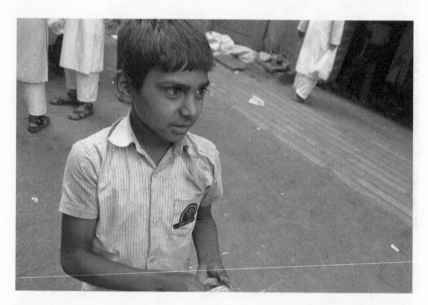

Madhu, a Hindu boy of the Harwal (also Heri or Heriyal) Dōm pariah caste, begging in Nizamuddin's alleys, 2015. His mother accompanied him on his rounds.

Notable is the abundant presence of Hindus who come to revere the
tomb (see Dhaul and Kamath 2006). Many neighborhood stakeholders and
residents, including beggars, remarked on charity to Hindus and Muslims
alike. The indiscriminate and unconditional charity Nizamuddin advocated
was inflected with an emphasis on outreach to Hindus, and to lower-status
ones in particular; a visit to the dargah is not at all incommensurate with
Hindu tradition or practice in Delhi, and the same is true of many other
Sufi shrines across the city and especially of the Dargah Sharif in Ajmer.
This is, furthermore, echoed in the tomb's eclectic and syncretic collection of ritual components: throughout the grounds are scattered trees with
candle-filled niches and tattered cloth tied on for good luck and hand-hewn
iron candelabras with ghee lamps burning in their various nodes.[13] Some
of these are staffed by faqir-type mendicants to whom a small donation is
given for the wish or the other auspicious benefits they might facilitate.[14]
      Nonetheless, distinctions between the variable meaning in local forms
of Hinduism and local forms of Islam of giving to beggars at a site where,

Women at the sanctum wall, Hazrat Nizamuddin's tomb

correspondingly, both beggars and donors may be Hindu or Muslim (or otherwise) are relevant here — alongside differences in the meaning of mendicancy itself in each. Such things structure, after all, the nature of the transaction between donor and beggar. Bornstein (2012) provides a rich account of the reasons in Delhi for giving (or not) from the donor's perspective, contrasting the Hindu practice of *dān* and the Muslim zakat. "Unlike zakat," she writes, Hindu "dān is not a reciprocal gift and does not create solidarity with fellow Hindus. In contrast, renunciation structures the practice [of] dān" (28). This suggests that at a place like Nizamuddin, the fate of beggars may be tied in the eyes of residents and pilgrims more cohesively to the fate of the place; it is a matter of wider communal concern. Indeed, this implies that the very act of giving to the poor helps to cement certain cultural or religious senses of self and experience, makes someone more broadly part of a Muslim interclass ecumene. Moreover, emphasizes Bornstein, dān must crucially occur at "the appropriate time . . . and the appropriate place" (28); thus beggars out of place (or out of time) are a matter of concern. Bornstein further shows quite engagingly the role of suspicion in the process of giving and the imperative of finding "suitable recipients"— a donor's calculus that is at work in force here in Nizamuddin among both Hindu and Muslim visitors.

For the runaways who circulate extensively in such spaces, I noted, birth religion seems not to be an issue, or rather it is actively rendered and framed as a nonissue. Indeed, many of the children in such space profess a kind of fundamental syncretism. Hanuman Mandir is, again, where I came to know Nabil, a Muslim by birth raised in an area with a nearly 100 percent Muslim population; Nabil said to me that "all the religions are one," that it didn't matter, God was God whether it was Bhagwan or Allah. I was surprised to hear almost precisely the same formulation at Kali Mandir as well, home to many Muslim children.

I referred here, fancifully, to a God of Urchins. But by using that phrase, I intended to highlight something special. What is it that might be understood as *unique,* as distinctive, about the "vagabond" child's place specifically in South Asian sacred space? Is there, we might ask, a proverbial God (or gods) of Urchins, emerging out of a *sant* and *nirguṇa bhakti* milieu? Is there a special place for such children in temple, shrine, and gurudwara? Might they be part of others' imaginaries of the space—even potentially part of its sacred quality (or themselves somehow sacred)? If yes, then why? This is not a question that can be definitively answered here, but it is one worth asking, underscoring, thinking about, and ultimately leaving unanswered. Is it something about the role of solo children in this part of the world, for example, or something about the religious traditions themselves and their accommodation of the poor and vulnerable? Certainly the beggar, for one, plays a distinctive and positively valued role in the deep history of South Asia; so too the orphan, the sage child who challenges death (as Nachiketas) or becomes a guru, and the child in flight from oppressors and cruel parents. The medieval words of those saints entombed in the shrines beloved to lawaaris children make frequent reference to the destitute child and the orphan and to the imperative of a service blind or perhaps compassionately open to communal identity. "Never refuse," wrote Khwaja Muinuddin Chishti, in his final discourse a month before his death in 1236, "to bless and help the needy, the widow, the orphan, if they come to your door."

As for the children themselves, they have a remarkable and remarkably conscious sense of the traditions they embody, occupy, and navigate and what their place in the spaces corresponding to those traditions means: what a *dargah* is (in general and to them and their compatriots), what a *mandir* is, who they are in those spaces simply as children in need (as they are legible to the public), what those spaces' denizens and pilgrims think of them (merely as children in need), and who they are in those spaces as who they *were*. In other words, on the last question, they have clear formulations of what it means to be someone at a Sikh gurudwara who has come from a Muslim milieu, or someone at a dargah who has roots in a low-caste Hindu milieu, among other such permutations, sometimes with even more specific specifications (low-caste Bengali Hindu, Bihari Muslim) as *meaning something*. And the indigenous sacred lurks in interesting ways in narratives of runaways' lives: to keep healthy, you should *always* eat your *tulsi* (holy basil), said one Hindu runaway from Madhya Pradesh at Nizamuddin Station to me. This was the same moment I describe elsewhere when I met several runaways from Ajmer itself, home to the tomb famed for harboring runaway children, who boasted about their variable connections to the shrine. When asked about intercommunal relations in such spaces, being a Muslim among Hindus or a Hindu among Sikhs—a charged topic in a country burdened with communal violence (see Blom Hansen 1999; van der Veer 1994; Menon 2010)—the children often boast specifically, as I observe above, about the oneness of God and how one should revere the dargah or temple, how religions are all one—sometimes while putting an arm around a compatriot of different religion. Sometimes, even more specifically, they comment on the patron (Sufi) saint or (Hindu) god who has helped them to the extent that she/he has, in a gesture of thanks, and often cross-communally.

This might potentially be seen as the articulation of a counterdiscourse to communalism: we're all here in this shitty situation, so we have to dispense with those animosities that others have time for. Indeed, this goes so far as to be etched onto the body: Hindu children bearing crucifixes as pendants or tattoos, Muslim kids with "Ram" carved on an arm.

In South Asia, such sacred spaces are, for kangaal, aawaara, lawaaris children, domains where a sense of "who it is that is like us"—a solidarity of pariah and outcast, and a lateral sense of who belongs in that social and physical space besides them—comes into resolution. Lepers and Dalits, dispossessed nomads and widows, orphans and runaways all find there a commonality of externality to one world, and internality to another. Indeed, such spaces are known as sanctuaries (in both senses) to and by children in flight even *before* they become "street children." Some indeed *start* their journeys with a circuit through such domains. Anjeel, a Hindu returnee from Roorkee, left home and before at length going to Delhi went right to the Golden Temple in Amritsar, where he served devotees, and then to the Hindu holy city of Haridwar. The sorts of sacred space frequented sooner or later by child runaways in South Asia are also sometimes, at least in the ideal, given their shape and meaning through a powerful and dialectical contrast to the profane world that such pariahs must otherwise occupy, the noisy, lecherous, hungry world of train station and dump and curb and *dhaba;* they are indeed in such ways constructed as *counter*spaces of a sort, contrastive spaces of refuge and retreat, special and peaceful sanctuaries imbued with magnetic, primal power, where all that filth and dealing not only can be left behind but is also—again in the ideal—not permitted to enter (see again Douglas 1966).[15] It is meant to stay at the door.

In contrast to home, furthermore, where children face pressure by virtue of their roles and personas, or organizations where they must register and be enumerated and face fixed trajectories, such sanctuaries, which exist largely outside the realm of the bureaucratic and of proper labor, afford anonymity and sometimes space or time to decide what's next; few questions are asked. This is not to say that they cannot or will not be exploited here; indeed they can and will be. But runaways identify something of value here, something to seek. They seem to know that in such shrines children are, on some level, for better or worse, part of the life of the space, known and expected there. It is as if the tomb or temple was waiting for them, ready to welcome them, the easiest to get to, and the best known, of possible initial or eventual destinations. And that is certainly not a *new* feature of sacred space in South Asia.

Such spaces may also be counterspaces not only in the provision of an *affective* contrast to the gritty world of the city but also in a socially and materially real sense in their provision of the possibility for *escape;* in this sense, they also hold the potential to become domains to which children may run from untenable labor situations and find some sanctuary (so to speak). Recall Ashok, who was brought to Delhi by what sounded like an informal trafficker who beat and confined him. Staying with this captor was not a possibility, he explains: "I ran away from him and came to the temple. In the temple I worked for someone to sell *diya* [sacred lanterns that burn butter]. . . . Aarti's mother really cared for me, the family liked me a lot. I stayed with them in the temple. Aarti's mother worked in the temple selling *prasad* [holy food]." Ashok says, then, the temple became forcefully redemptive in a number of ways, including in the refuge it gave him from an exploitive labor situation and in the immersive participation it allowed him in what appear to be aesthetically appealing elements of ritual life, and even a kin structure that was not his own. Compelling narratives by children about such sites and the saints, gurus, or gods to whom the sites are devoted also suggest that they bestow good on a soul by association—they might be seen especially as imbuing good on a soul or a period that feels marred or tarnished, or by contrast to a city and its spaces that feel marred and tarnished. There is nothing else quite like these spaces for these children, nor indeed anything else like these children for these spaces: a unique cultural complementarity emerges in the dialectic between person and space here.

The question that intrigues me here is whether sacred spaces at which runaways seek refuge are themselves *consciously* in some way not just spaces that contrast the world of misery beyond their gates but counter*institutions*—that their role (but not their existence) somehow forms a dialectic either empirically or at least from some particular perspective with the formal institution of child rescue, salvation, and reform. I believe they are and that they do. From my point of view, they provide similar services but in a radically different modality that presents a fundamentally different, and possibly opposed, structure. And for the runaways themselves, again, I perceive that spaces like Nizamuddin exist in contrast to other modern institutions, organizations, NGOs—here, for one thing, no questions are asked, no lists made, no performance assessed, and no strategy for self-repair espoused, other than a display of devotion.

But if there's anything to this hypothesis on counterinstitutionality, then counter to what? Capital? Modernity as such? Indeed, to the public, to the municipal authorities, to the executors of spatial cleansing, these places, like the people they help care for, are also considered a taint on urban space (Ghertner 2015). That marking itself from without indicates the conviction in the threat that such spaces pose to an aesthetic modernity. From within, too: these stand as an *alternative* to the corrective and reformative, a something else altogether that demands neither computer literacy nor a return home.

Labor and Salvation: Why Imperial Ghosts Haunt Postcolonial Charity

I would like to underscore the thin boundaries both *historically* between colonial campaigns of child reform and contemporary efforts at child aid and *synchronically* between humanitarian worlds of child protection and carceral worlds of child detention in the contemporary moment. The two boundaries are furthermore potentially on some level basically the same, if arguments are taken seriously that colonial modes of child incarceration and detention under the guise of charity in fact constitute the historical roots of contemporary modes of child welfare and charity.

It is essential to underscore, in this context, that not all efforts at addressing the undeniable and immediately visible misery faced by countless children in India are the same; indeed some organizations, advocates, and movements intentionally critique or resist the dominant narratives, ideologies, and practices of child-saving corrective charity, some of it lingering from colonial child rescue, that pervades the language of many institutions, whether private or state. These could be said to constitute a small class of "counterinstitution" consciously committed to a radical approach to childhood suffering. Some of these might be situated in radical politics more broadly applied, as embodied in the emphases of actors like Vandana Shiva, Sunderlal Bahuguna, or Arundhati Roy, who speaks explicitly (2014) of the "NGO-ization of resistance," or movements like Shikshantar, which seeks to transform the shape of education in India. Of note in the street children's and runaways' domain is the Mumbai organization Sadak Chaap (the name refers to a term used by street children for themselves, meaning "the stamp of the street" (i.e., those who bear or wear that stamp)), whose Latin America-inspired "process constantly challenges the Juvenile Justice system which does not work. And whose remanding to state custody is one of the most archaic and senseless process in which young children feel criminal and in 'Jail' "; Sadak Chaap encourages children "to form their own loose federation," and works to create a cadre of children who might ultimately become leaders "in the larger movements of the poor."

Satadru Sen (2005) deftly illuminates the origins, culture, and shape of the imperial Indian child reformatory, underscoring its techniques of control and its intimacy with projects of rule, colonial and national alike. He does much to reveal the conflation of discipline, reform, and aid, the intersection of detention and charity, in the history of Indian institutions for children. Charity and reform were indeed at the time fundamentally inseparable. For this argument, Sen offers us a glimpse into the intimacy of a coercive project, one ostensibly about "rescue" and salvation. Sen's research also underscores the inextricable interdigitation of morality with modernity (including through an incisive analysis of predicaments surrounding flogging). Even the basic lexicon of the whole affair—"reformatory," "delinquent" (114)—denoted such imperatives, urges, and interests. Indeed, Lord Napier, for example, "drew an explicit parallel between Indian and English 'strays,' seeing both as unsocialized, non-nurtured tabulae rasae upon which the child-saver might inscribe *morality and modernity*" (Sen 2005: 63, emphasis mine).

If the line is thin between colonial and current modes of child reform, then the line must also be thin between such institutions in the present and the missionary urge that underlay such reform in the past. Indeed, I find in the archive itself the same Lord Napier pointing precisely to that: in the *Abstract of the Proceedings of the 1874 Council of the Governor General of India*, for example, he declaims:

> The absence of moral and religious teaching in a juvenile reformatory would be undoubtedly a serious want. . . . It must be conceded that moral teaching lose[s] as much of its force of will not associate[d] with the influences of Christianity. . . . I am not prepared to admit that all direct moral teaching must be banished from an Indian reformatory . . . but, granting that I am mistaken, and that direct moral teaching is impracticable in an Indian reformatory, there still remains the invaluable force of indirect, insensible, practical discipline—perhaps more powerful than preaching and teaching on the human heart— the habits and lessons of obedience, cleanliness, punctuality, industry and economy which may all conduct the criminal child in India, as well as in England, to a better stage of life. (India Imperial Legislative Council 1875: 168)

This sounds in part like a conflation of evangelism and child reform, and in part a conflation of charity with discipline (to/of/for modernity), but there is no mutual exclusivity in such a tension. As Comaroff and Comaroff (1992) argue, missionary evangelism of the Tswana and the inculcation of the "native" population with precepts of proper, industrious, and ordered labor were inseparable. An equivalency was built between the yoking of local labor to European markets and the joining of the soul through work to heaven. Work was itself morality in this teaching. And indeed, farming, they write, "would cultivate the worker as he would cultivate the land. The production of new crops and the production of a new kind of selfhood went together in the Evangelical imagination" and the cultivated garden produced in the process "was an icon of the civilizing mission at large"—its labor, its "material individualism," its moralities of private property, and its correction of "idleness" (Comaroff and Comaroff 1992: 246–47).

The description of educational efforts at the Hazaribagh and Alipore Reformatories in the 1902–3 *Report on the Administration of Bengal* (published 1904) are striking in this regard. Hazaribagh, which was for rural delinquents, taught as part of the reform project "cultivation and market gardening" while Alipore, aimed at urban miscreants, "instructed in various industries" (125). Nonetheless "the figures with regards to occupations followed by boys after release [were] still very disappointing" (125). Earlier, in the report for 1880, Larymore (1881) explains, "Many of the inmates of the Reformatory are very young boys, many of them with sentences long enough to suffer from the effects of bad example from the bigger boys, who have shown themselves to be incorrigible, and yet *far too short to ensure any lasting effect from the moral and industrial training* under which they are brought" (3, emphasis mine). Elsewhere he urges that inmates "sent to a Reformatory School should be of such an age as to be at once practically brought under the influence of the moral and industrial training the school offers. Between the ages of eleven and fourteen years is, I believe, the most favorable period for juvenile offenders to be sent to a Reformatory" (4). But Larymore laments the inexorable inherency of the delinquency: "often," he writes, "instead of education and moral training we have unexampled obstinacy and perverseness, coupled with a preference for all that is evil and degrading" (4).

This is tied to an essential conviction about the natal moral corruption of the non-Christian society, a conviction that went so far as to deem certain castes and classes, as discussed in the last chapter, "criminal" by birth (or, post-1952, "habitual offenders"; see Radhakrishna 2001; Viswanathan 2014). Such people, as I shall show shortly in greater detail, were seen as unteachable, ungovernable, unsocializable to capital, labor, and industry. Culture itself was the object of reform. I raise this because I feel very firmly that the question of uncorrectable, aberrant culture remains in the governing of runaways: the children's culture does not know how to do things right. Even within the larger field of play, however, in the missionary and colonizing efforts, culture—at scales as large as "native" worldwide, as Indian "native" macroregionally, or as any community subregionally, such as "Muslim" or "Thuggee" or "Nath"—was seen as an object of correction. Subjects had to be swayed from the taint of birth that imprinted itself in belief, practice, and society to the right and rational way of Anglo-Christian Truth.

My long forage in those crumbling (formerly India Office) records at the British Library yielded some other interesting revelations on the intertwining of Christian salvation and child reform in various periods of colonial India. Hobbes's handwritten "Scenes in the Cities and Wilds of Hindostan" recounts the story of a Mrs. Wilson who

> In the year 1832 . . . took charge of a number of orphan children whose parents had perished by disease and want during the famine which had that year assaulted the country. She determined to give them a Christian education, apart from all heathen connections, and with this view erected by subscription a spacious building on the banks of the Ganges, about seven miles from Calcutta, which was opened in 1836 with nearly a hundred orphans. She also erected a school near the Refuges, capable of accommodating about 400 children, which was opened in 1838, subsequently a Mission House for a resident clerical Superintendent, and an elegant and substantial Church. The Church, the School, and the Mission Houses, have since been made over to the Church Missionary Society; and the Bishop and Archdeacon of Calcutta have taken charge of the Orphan Refuge. (1852: 122n)[16]

Elsewhere, in the *Calcutta Christian Observer* (1837), it is presaged that the "Berhampore Orphan Asylum, composed of industrious moral Christians, will become a true *oasis* in the desert of ignorance, bigotry, and superstition; and will practically *shew* to the surrounding multitudes, the positive blessedness of Christianity" (678)—in part by training indigenous missionaries. This was to be—and was—accomplished by rigorous and highly time-regimented work, including agricultural work. Here the orphan was to be "industriously employed and earn his own support, without idleness or dependence, at the same time as he is learning that which is able to save his soul and make him at once a happy, a good, and a useful member of society. . . . Idlers and impostors will not long remain in an institution where the apostolic maxim is strictly a rule — 'if any will not work, neither shall he eat'" (678).

Such sentiments, urges, and imperatives were toned down and more subtle but still alive and well more than a century later, around the end of World War II, as the independence movement began to crescendo (as I argue they are now). I found at the archive in London a handwritten India travel diary, for example, of a Margaret C. Godley, a social worker and "women's rights activist" of minor note, with the following entry, containing intimations of more modern and even progressive sensibilities on child welfare but also undertones both that Indian society, savage as it is, does not know how to treat children as well as the beleaguered veteran "Mrs. Davis," or her allies in the juvenile court, and that the street is a place from which — and begging an occupation from which — racialized children must be rescued. (I endeavored to preserve orthographic details, but question marks here denote difficult-to-decipher handwriting):

> Poona 27.9.1945
> Visit to Mrs. H.K. Davis, Resource[?]-Worker at Thatila/Hatila
> Seva Handel [?], Terandavana
> Deccay Gymkhana, Poona
> An Englishwoman came out of the bungalow to greet me. She
> was short, pathetically thin (although obviously skinny), had
> a brown face covered in fine wrinkles, and wore her hair in a
> neat bob. She was dressed in a cheap and rather dirty cotton
> frock and wore a pair of old leather Indian sandals. Her voice
> was cultivated, her laugh spontaneous, and her eyes had a re-
> freshing twinkle in them. This was Mrs. Davis, the woman who
> had been struggling since 1924, in Bombay and then in Poona,
> for the well-being of women and children. She had lived alone,
> in this sordid bungalow, devoid of furniture and the amenities
> of life, alone, unnoticed in a world geared only for war and al-
> most unbefriended, for nearly five years, engaged in a struggle
> against the exploitation of children, and the enforced prosti-

tution of women. There must have been little that she did not
know about underground life and crime in Poona. She said
that her staunchest helpers and others were two members of
the Juvenile Court.

    Two little brown faces were pointed out to me among a
crowd of children of all ages and all creeds. These two she herself had rescued from a street in Poona, where they had been
found, sprawling and starving, in the clutch of a beggar.[17]

While perhaps suggesting something about public opinion on child suffering in its time, this reveals more about the emergence of modern mores in
child welfare than the persistence of missionary ones, or indeed perhaps
the meeting point of those threads. Nonetheless, it reveals certain implicit
ideological underpinnings of the emergent English interest in saving Indian
children—race, for one, is certainly quietly present here, as is "rescue."
One wonders: was the "beggar" from whose clutch the Englishwoman had
"rescued" the two possessors of the "little brown faces" kin to the children?
It might well not be the case that she or he was indeed kin. But it might be.
And in such cases, one wonders what constitutes child salvation.

    Back to incorrigibility—culture as cause for detention, and the place
of tribal identity in decreeing delinquency. When it came to those tribes
in reformatories—juveniles placed in detention by virtue of their heritage
alone—the operant question was whether it was justified to separate such
children from their parents, whether their culture mandated such a separation, and thus fundamentally also whether England knew better than their
culture what childhood itself was. This manifested itself in public debates
in interesting ways. When, for example, the imperial government of India
objected to the proposition that children should be removed from their parents, Lord Napier protested vigorously, arguing for removal:

> It is considered in Europe that one of the greatest benefits which
> can be extended to the juvenile criminal in most instances is
> to be separated from early associations and from parents who
> are often his instructors in crime, and who live by the fruits of
> his iniquitous industry. Crime is hereditary, it goes down from

generation to generation, and the reformatory is the beneficent power that breaks the chain of transmitted guilt which weighs upon a family once embarked upon this fatal course. I cannot state from personal knowledge to what extent criminality is propagated by parental teaching in this presidency. The ancient murdering and thieving castes are no doubt much broken; still, in the country of Thugs, Maravers, Dundassies, Pullers and Kullers, a hereditary character must be more deeply stamped on crime than in England. (India Imperial Legislative Council 1875: 168)

What are the methods and antecedents for the correction of hereditary sin in the intersection between evangelism and secular colonial and postcolonial structures of child welfare? Erasure of culture, expulsion of population, and the rending of families as was seen in indigenous parts of the Americas and Australia—or indeed of Roma or Jews throughout history. One need only hear the tale of the Jewish children taken to São Tomé or the Indigenous Australians to boarding schools or white homes, away from their families, to be convinced of that.

The law on criminal tribes (An Act for the Registration of Criminal Tribes and Eunuchs)—and on children to be detained because of their hereditary community—reads as follows:

ACT No. XXVII of 1871
Part I.
Criminal Tribes

2. If the local government has reason to believe that any tribe, gang or class of persons is addicted to the systematic commission of non-bailable offenses, it may report the case to the Governor General in Council, and may request his permission to declare such tribe, gang or class to be a criminal tribe.

3. The report shall state reasons why such tribe, gang or class is considered to be addicted to the systematic commission of non-bailable offenses and, as far as possible, the nature and the circumstances of the offenses in which the members of the tribe

are supposed to have been concerned; and shall describe the manner in which it is proposed that this tribe, gang or class earn its living when the provisions hereinafter contained have been applied to it.

4. If such tribe, gang or class has no fixed place of residence, the report shall state whether such tribe, gang or class is to be settled under the provisions hereinafter contained. . . .

13. Any tribe, gang or class which has been declared to be criminal, and which has no fixed place of residence, may be settled in a place of residence prescribed by the Local Government.

14. Any tribe, gang or class, which has been declared to be criminal, or any part thereof, may, by order of the Local Government, be removed to any other place of residence.

17. The Local Government may, with the sanction of the Governor General in Council, place any tribe, gang or class, which has been declared to be criminal, or any part thereof, in a reformatory settlement.

17A. (1) The local government may establish and maintain reformatory settlements for children and may separate and remove from their parents and place in such a reformatory settlement the children of the registered members of any tribe, gang or class which has been declared to be criminal. . . .

(3) The superintendent of reformatory settlements for children shall be deemed to be the guardian within the meaning of Act No. XIX of 1850 (concerning the binding of apprentices), of every child detained in such settlement; and such superintendent may, if he shall think fit, and subject to any rules that the Local Government may make in this behalf, apprentice such child under the provisions of the aforesaid Act.

Explanation—the term "children" in the section includes all persons under the age of 18 and above the age of four years.

The last provision, section 17A, clause 3, on apprenticing, is of particular note in light of the larger argument here on breaking, corralling, and cultivating children, particularly children from dispensable classes, for work. It drew, itself, on an earlier law:

Act No. XIX of 1850
Concerning the Binding of Apprentices
For better enabling children, and especially orphans and poor children brought up by charity, to learn trades, crafts, and employments, by which, when they come to full age, they may gain a livelihood. . . .
4. An orphan or poor child, brought up by any public charity, may be bound apprentices by the governors, managers or directors thereof, as his or her guardian for this purpose.

With the provision for apprenticing, and especially in the intersection of the 1871 and 1850 laws, which permitted the reformatory headmaster to serve as the child's guardian to induce a situation of an apprenticeship that was essentially bondage if not enslavement, the detained tribal child in the reformatory was like the missionary's subject, rendered available at the very least for imperial labor. This was of course part of a larger complex of Indian indenture, debt servitude, and peonage fostered by British institutions that gained particular vigor in the wake of the British abolition of slavery. But it wasn't new. Marx (1867), in *Capital* (vol. 1), notes the long presence of such forms in England, citing an edict of Edward VI: "All persons have the right to take away the *children of the vagabonds* and to keep them as apprentices, the young men until the 24th year, the girls until the 20th. If they run away, they are to become up to this age the slaves of their masters, who can put them in irons, whip them, &c., if they like. Every master may put an iron ring round the neck, arms or legs of his slave, by which to know him more easily and to be more certain of him" (VIII.XXVIII.4).

The larger puissance of the Criminal Tribes Act was rooted in the powers it gave the governor to publish the names of criminal tribes in a gazette and to call upon all members of such criminal tribes to appear to register themselves. If they did not or they dissimulated, it was a prosecutable crime. This naming rendered criminal tribe children subject to abduction by the state and adoption by reformatories. Within villages of "criminal tribes" there were roll calls and passes to license movement. Village headmen and village watchmen were tasked with policing their kin and neighbors and reporting violations to the police. If they failed, they were themselves subject to punishment. Punishments for noncompliance included whipping, fines, and imprisonment. And there was a three-strikes rule — break the law thrice and go to jail for life. The 1871 law remained in effect until 1952, when it was replaced by the Habitual Offenders Act, which "denotified" the tribes (making them DNTs, Denotified Tribes, but still a special category). The Denotified Tribes — "Habitual Offenders," legally speaking — emerged from their near century of legal separation largely far from prior homelands, without their traditional means of subsistence, and staggeringly poor.

It is of note that such narratives and sensibilities on culture as generational criminality must have been informed at least in part by constructions of "Gypsies" — Romani people — at home in Britain, the locally known "hereditary criminals" par excellence, according to British lore. Indeed, the treatment of Roma followed these very contours of culture correction, where possible (though a structure for the separation of Romani children from their parents was never systematically established in England). All this belies a troubling conviction in culture as grounds for detention and for the "reform" — the *fixing* — of children. I am impelled to speculate that a tacit piece of such a view still guides the insistence on repairing runaway children, and restoring them to their places of origin. The unspoken ideology may be in part that those cultures do not themselves know how to care for, raise, or keep their children, and thus the child must be corrected and so must the culture. In certain cases it is deemed appropriate that the child be kept from her family, her place, and thus her culture. Among urban residents, and often even among NGO workers, discourses that deem the children's places, cultures, and parents as unfit, and in dereliction of their duties, abound and persist. Often the case is made that the real culture of the parents themselves must be rectified, that it is Culture Gone Wrong, rather than structures that underlie it. This is a problem described by Scheper-Hughes (1992), Donna Goldstein (1998), and Montgomery (2000) about the urge to blame parents rather than poverty, capital, history, or states for poor children's suffering. Thousands of miles away from India, indeed, this is the view that guides the state's sense, once again, of entitlement in South Dakota when it removes thousands of native children from their families and places them in foster care.

It is far from disappeared, this curious and sad constellation of people, law, and space, and indeed it is not at all done, in part because of the terrible aftermath of the Criminal Tribes Act and its institutional imperatives. In a recent trip, in 2015, I found that Phasē Pardhi people, members of an 1871-stipulated Criminal Tribe (later "Habitual Offenders," circa 1952, or "Denotified Tribes"), were residing en masse — with numbers perhaps in the hundreds — in the intersections and sidewalks around Connaught Place's Hanuman Mandir. Many of these were street-working children; some were orphans. The Pardhi were in fact before their designation as hereditary criminals forest-dwelling people of eastern Maharashtra, near Aurangabad and Nagpur, primarily hunters, whose lives were more like those of other groups now clustered under the designation Adivasi, aboriginal. Gradual urbanization and deforestation pushed them to the edges of towns where, in British reckonings, they began to engage in "criminal" lifeways of petty crime and theft. When persecution, eviction, and exclusion made their prior lifeways and domains unlivable for them, they began to migrate in force to inhabit Delhi's curbs and medians.

Child reform in India started largely as an element of missionary campaigns. Missionary work was, as has been forcefully argued, in part in the service of capital and its moralities of modernity and labor; culture was its object, erasure by reform and conversion, or reform as conversion, its technique. Even official and secular humanitarian efforts intersected with the missionary project, and all of it revolved around the motif of a correction ("reform," "rehabilitation") to be achieved through labor. That correction is consonant with later efforts at "mainstreaming" and — still — "rehabilitation" through socialization to class norms and norms of modernity, again through work. And from there we can come full circle — if child reform can sometimes be in the service of at least norms of modernity and capital, then, if we listen to the Comaroffs, its intimacy with missionary pasts cements this even further to present a kind of complex of socialization of children to the right way of being, a way of being guided by knowing

A homeless indigenous Pardhi boy, Connaught Place, 2015

one's proper place in regimes and cosmologies of consumption, accumulation, and production as embodied in behavior and comportment and a certain normalcy, where the end, the goal of the missionary effort, is conversion to the salvations and truths of modernity and global capital itself, not such a grand break from the past as it might seem.

The thin line between current and colonial is in the end then also the thin line between charitable and carceral, which exists in both the contemporary and historical dimensions, as colonial child-reform structures—and laws—persist much as they were. The same laws, in this period, may even just as easily send children in similar situations to a "remand home" as to a humanitarian "shelter" in partnership with the government. And thus all these boundaries—charitable and carceral, missionary or rehabilitative, colonial and contemporary, purgative (in the context of street evictions and cleansing) and humanitarian (in the context of institutionalization), psychiatrically normalizing and socializing to capital—are on some level the *same* boundary.

SEVEN

Concluding Thoughts, Final Words, and Big Pictures

Indian Runaways in Larger Frames

Who, Why, What, and So What: The *Times of India*

From time to time the Delhi Department of Woman and Child Welfare runs a full-page notice in the *Times of India* with a banner identifying it as a list of "Children Separated from Their Families." In a notice from June 1, 2007, below this banner appear photographs of twenty-two boys; next to each photo, a phenotypic description, a place of putative origin, a nickname, and some auxiliary information:

> THIS BOY RAKESH ALIAS TULSI S/O SHRI MAGAN, DABRI
> WAS ADMITTED IN THE ORGANIZATION ON 26.8.2004.
> AT THE TIME OF ADMISSION HIS AGE WAS AROUND EIGHT
> YEARS, HEIGHT 4 FEET 5 INCH, COMPLEXION DUSKY TO
> BLACKISH, BODY MEDIUM, FACE OVAL. IDENTIFICATION —
> A CUT ON LEFT LEG. ADDRESS AS TOLD BY THE BOY: VILLAGE
> BANTA NAGAR, DISTT. TATA NAGAR, BIHAR.[1]
>
> THIS BOY JITENDRA KUMAR ALIAS BAJRANGI S/O SHRI
> ARJUN SINGH WAS AROUND 8 YEARS OLD AT THE TIME OF AD-
> MISSION IN THE ORGANIZATION, COMPLEXION BLACK, FACE
> OVAL, PRESENT HEIGHT 4 FEET 6 INCH, IDENTIFICATION
> MARK — SCAR OF BURN NEAR RIGHT EYE. THE ADDRESS AS
> TOLD BY THE CHILD, VILLAGE JHUGGI, TWO K.M. AWAY FROM
> LUCKNOW, TEMPLE OF BAJRANG BALI, LUCKNOW, U.P.[2] THE

BOY WAS BROUGHT BY POLICE STATION NEW DELHI RAILWAY
STATION ON 8.10.2003.

A close read of these little synopses of terrible and complex situations presented in the newspaper encapsulates much about what this book has been
concerned with—immediately one sees children who have long been in
custody, of similar ages at the time of admission, with almost no information whatsoever on what happened to the children, why they were separated from their family, and how they got to Delhi. This book has been,
throughout, about those questions: how, why, and what. The unwritten
history behind those gaps in the text, the lacunae determining the shape of

"Lost Child," Kali Mandir temple passage wall

those short and partial biographies. Those stories, in turn, point to a bigger who, why, what, and indeed *so* what, to larger formations that the stories both *obscure* and *are,* and to complexities in the telling of the stories themselves, both here, by me, and there, by them.

Who Am I to Tell Your Story?
On Whether and How We Write Suffering

The intent of Adorno's oft-quoted assertion, largely in response to Celan's (1948) *Todesfugue,* that "to write poetry after Auschwitz is barbaric" (1982: 34) is ill understood and widely decontextualized. In an elaboration, Adorno suggested that as "art is unable either to experience or to sublimate suffering, Celan's poems articulate unspeakable horror by being silent, thus turning their truth content into a negative quality" (1984: 443–44). Later, indeed, Adorno rescinded, a bit, writing that "perennial suffering has as much right to expression as a tortured man has to scream; hence it may have been wrong to say that after Auschwitz you could no longer write poems" (1973: 362). Nonetheless, the original claim has been mobilized in the development of a set of essential scholarly questions not only on writing about the deepest suffering in general, but also on who can write about another's collective suffering, on the degree to which it is either appropriate or an appropriation, on the matter of stealing voice: Who am I to tell the story of someone else, and particularly of an oppressed someone else? What if the type of person I am is at some level implicated in that other's oppression? Certainly the streets of Delhi are no Auschwitz, and this might not be art. Nonetheless, this is a book largely about misery and mass suffering, a suffering tied to terrible lives and deaths. In the face of that, what ethical imperatives for the ethnographer or author?

In *Charred Lullabies,* Daniel (1996) engages with this nexus of Adorno and Celan to try to make sense of a landscape of violence in Sri Lanka, and in particular to wrestle with how to capture in—or reduce to—narrative *life,* when life has turned fully to terror, or horror. "I lurch," he writes, "towards the rustle of hope, wherever it may be heard or found. But in so doing I run the risk of arrogating event to word, assimilating brute fact to cultural reality, translating event into language, wringing out of the cries of the boy consumed by the inferno an alibi for his murderers" (211).

Nancy Scheper-Hughes asks, likewise, Who am I to speak for this misery? in her seminal *Death without Weeping: The Violence of Everyday Life in Brazil* (1992). *Death without Weeping* is a chronicle, nearly epic in breadth, length, and style, of the life of a favela in Brazil's poverty-stricken sugar zone. It is about, among other things, the historical nature of what at first appears to be essentially the maternal abandonment of infants who look like they are not going to thrive, and more fundamentally about a suffering patterned by history and its predations. Such scenes evince in authors and viewer alike the question Who am I to be spinning this tale? or Who am I to be watching it? Even James Agee faced that predicament as he set out to write *Let Us Now Praise Famous Men* (1941), decades before (temporally) and worlds away (conceptually and institutionally) from where these questions entered currency:

> It seems to me curious, not to say obscene and thoroughly ter-
> rifying, that it could occur to an association of human beings
> drawn together through need and chance and for profit into a
> company . . . to pry intimately into the lives of an undefended
> . . . group of human beings, an ignorant and helpless rural
> family, for the purpose of parading the nakedness, disadvantage
> and humiliation of these lives before another group of human
> beings, in the name of science, of "honest journalism" (what-
> ever that paradox may mean), of humanity, of social fearless-
> ness, for money, and for a reputation for crusading and for un-
> bias which, when skillfully enough qualified, is exchangeable at
> any bank for money . . . and that these people could be capable
> of meditating this prospect without the slightest doubt of their
> qualification to do an "honest" piece of work, and with a con-
> science better than clear, and in the virtual certitude of almost
> unanimous public approval. (7)

In classes I teach on war, or street children, or other forms of collective suffering my students ask this, wondering about "misery porn." Are we just consuming other people's suffering as aesthetic object? What does it mean and what does it do to be sitting here pounding these images into our head — and yet still *just sitting here?* What would it mean, I usually reply, *not* to? Not to record, not to recount, not to see — to avert our gaze and stay our pen. What is the alternative? Isn't it a debt, I ask them, if someone else has to live it, for us at least to *know* it, if not to enter the fray and renounce some of our comforts to join others' struggles? Some of them, invariably, find solace in the easy answer: it is better to live the lives we have and not engage, to be comfortable. The critique of the fraught nature of cultural representation and the fetish of suffering indeed seems to make it *even more okay* to be comfortable. This is the "distant suffering" — bolstered by the media — that Boltanski (1999) unpacks, and that can create "ever more robust cultures of denial" (Wilkinson and Kleinman 2016: 105). At last, we can watch TV and sip our wine and not worry, because that worrying would itself be unjust, an abuse of our power. It is surprisingly easy for the student to come to this conclusion, based on his or her understanding of the salient debates. And so, unwittingly, viewers' sense of the justification of privilege, of deserving it and not needing to engage, can entrench itself ever deeper when faced with the idea of the pointlessness of viewing suffering and the imbalance inhering in the directionality of the viewing.

But Scheper-Hughes finds the notion of ethnography as "unwarranted intrusions into the lives of vulnerable, threatened peoples" (1992: 27) and as "a hostile act that reduces our 'subjects' to mere 'objects' of our discriminating, incriminating scientific gaze" (27-28) — a notion she describes as inspired by Foucault — highly problematic, if not paralytic. Scheper-Hughes provides — for me, at least — a meaningful alternative vision (or, if it suits you better to think of it this way, a justification) of the potential power of fieldwork, indeed of fieldwork as *a work* of value in and of itself.

I think of some of the subjects of this book for whom anthropology is not a hostile gaze but rather an opportunity to tell a part of their life story. . . . I believe there is still a role for the ethnographer-writer in giving voice, as best she can, to those who have been silenced. . . . I believe there is still value in attempting to "speak truth to power." I recall how my Alto friends grabbed and pushed and pulled, jostling for attention, saying "Don't forget me; I want my turn to speak. That one has had your attention long enough!" Or saying, "*Tá vendo? Tá ouvindo?*"—Are you listening, really understanding me?" . . . Or "Write that down in your notes, now. I don't want you to forget it." Seeing, listening, touching, recording, can be, if done with care and sensitivity, acts of fraternity and sisterhood, acts of solidarity. Above all, they are acts of recognition. Not to look, not to touch, not to record, can be the hostile act, the act of indifference and of turning away. (28)

I think, one day, of a boy at the edge of New Delhi Railway Station who asked me what I was doing. When I told him, he replied, clearly incredulous, "Wait, you want to tell *my* story?" Not to make the act a heroic one. But there is something to having your story told. Scheper-Hughes does something for the ethnographer's doubt in this landscape, but also perhaps for the ethnographic subject's. She continues:

> There is another way to think about fieldwork. I am taken with
> a particular image of the modern ethnographer, one borrowed
> from John Berger. . . . The clerk, or the "keeper," of the records
> is the one who listens, observes, records, and tries to interpret
> human lives, as does the traditional country doctor. The clerk
> can be counted on to remember the events in the personal lives
> and in the life history of the parish. . . . She is their genealogist.
> . . . She can readily call to mind the fragile web of human relations that binds people together into a collectivity. . . . This is
> the image I suggest for the ethnographer-anthropologist . . . a
> keeper of the records, a minor historian of the ordinary lives of
> people often presumed to have no history. (1992: 29)

The key, it seems, and the test, is that the people involved see it that way as well. The question is, of course, not a new one in anthropology. From Rabinow (1977) to Rosaldo (1989) to Crapanzano (1980), ethnographers have been asking themselves, or really encouraging other ethnographers to ask themselves, What am I doing here? and Why should I do this? and, again, Who am I to . . . ? — to speak for others, to define someone else's suffering, to represent it to others, to profit, whether professionally or financially, from that representation, and to bring it to a space where others might enjoy looking at it? Why even go? Why ask? Why then write? Can something generative be gained from the documentation of others' suffering?

These questions are particularly potent in what Ortner (2016) calls the turn to or triumph of "dark anthropology," which is "anthropology that focuses on the harsh dimensions of social life (power, domination, inequality, and oppression), as well as on the subjective experience of these dimensions in the form of depression and hopelessness" (47) and that "emphasizes the harsh and brutal dimensions of human experience, and the structural and historical conditions that produce them" (49). Ortner rightly points out that all this has "not gone uncontested, and indeed has provoked strong reactions in some quarters" (58). Of course, as she notes, the specters of Marx and Foucault and Eric Wolf—adorned additionally by Asad, Said, and other analysts of the colonial—haunt and pervade this entire discussion. Tobias Kelly (2013), whose comments she cites, is concerned with the possibility "that the 'suffering slot' is now the rhetorical justification for much modern anthropology" and laments that

> book after book has described, in great ethnographic detail, the multiple ways the humans can be horrible to other humans, and the many forms that the experience of such horrors can take. Some of the work in this vein has been original and illuminating. It has shed important light on forms of domination, violence, and inequality. However, at the same time, some of us have also had a nagging feeling that, at times, there has been a little too much self-righteous enjoyment in all these descriptions of misery, without asking ourselves what role they serve. At what point does an ethnography of suffering turn into a voyeuristic quasi-pornography? What is the point in yet another description of the capacity of humans to feel pain and suffer? Does it really

make anyone suffer less, and do most of these academic publications really play any role, directly or indirectly, in the fight
against suffering and inequality? Even if our concerns are entirely academic, do the dozens and dozens of articles setting out
ever new and fine-grained descriptions of the horrible things
that many people have been through, add anything significant
to anthropological knowledge, whatever that might be? Surely
there is more to the discipline than the analysis of quite how bad
life can be? Such questions are urgent and important, and have
troubled many an anthropologist. Yet, the desire for a less miserable disciplinary focus also comes with anxieties of its own.
Is it possible to focus on the seemingly more positive aspects of
human life . . . ? There is a need to avoid simply providing descriptions of the harsh lives that many people live, without denying that many people do live such lives. (Kelly 2013: 213–214)

"Voyeuristic quasi-pornography" (which Ortner sees as a dismissal
of the works in question, incidentally) is exactly what those class discussions are about. But to me, it seems a little much. It seems to me to obviate
anthropology's potential to speak truth to power. Even if Kelly is right, is
a happy ethnography of positive fun—to be hyperbolic in order to make
the point—the antidote to it? What if a shantytown resident is proud to
hold the book that is written about her efforts to give her neighbors legal
protection against police raids? What if an ethnography uncovers a corpo-
ration's dumping of toxic waste into a slum, leading to litigation? What if
it is cited as justification for an ERC or Gates or NIH proposal calling for
vaccines, or condoms, or cancer screenings, and the study is funded, and
successful, leading to better or longer lives? Should such things not be the
pressing domain of ethnography? It seems to me we cannot be so terribly
cynical about illuminating suffering. And yet we also should spend most
of our time in illuminating suffering thinking about how that illumination
might alleviate it to the tune of the suffering people's own wishes. And we
should not assume that what we think of as suffering is necessarily always
what other people do. Our vision must meet the vision of the others we are
talking with, and about, and who are talking through our work.

And it must aim to transform oppressive structures. Philippe Bourgois (1996), in an article that seeks in part to "confront anthropology," and
that, to me, echoes some of Scheper-Hughes's forceful mandate (and coun-
ters Kelly's), writes that "ethnography's tremendous potential for initiating
contradictory dialogues that violate cross-class and interracial taboos in our
home environments remains mostly untapped," and though he frets over
the "inherent pornography of violence that automatically engulfs any pre-
sentation of the details of extreme social suffering in the United States," he
asserts, quite forcefully, and in affirmation of the project here, that "understanding and representing these problems offers more than an intellectual
exercise for anthropology: It is an urgent political challenge" (256). Paul
Farmer, more recently, in his Malinowski Lecture, insists that "anyone can
learn about the suffering of others, including that caused by serious illness
or injury. Anyone can stop and listen to the stories. . . . *Anyone can apply
not only knowledge, but also compassion and pragmatic solidarity, to lessen
the suffering about which we so often write.* It's a question . . . of epistemological humility and the cultivation of a humanitarian sentiment not to be
mocked" (2016: 286–87, emphasis added). And yet, it is, among some of
these other critics, being mocked. Also, inhering in Farmer's *anyone can*
is an implicit *everyone should,* above all the ethnographers who set out to
document and listen and have a choice in that matter, in what falls within
their selective ethnographic and then writerly scope. We must apply to
these choices this ethic of attention to critically urgent and shared human
concerns. The benefit might be for the subjects in question, or for the public: "being positioned as a witness [can] serve as a spur for the development
of social consciousness and as a provocation to engage in the pursuit of
social understanding" (Wilkinson and Kleinman 2016: 20).

But for some, the sort of revolutionary charge Bourgois endorses,
the ethnographic mandate Farmer encourages, may not be enough, may
still be fraught, laden with the baggage of centuries of twain legacies of
privilege and subjugation, or it may, rather, be too much. Eve Tuck (2009),
in a long open letter in *Harvard Educational Review,* urges a stance of
critical caution and ethical concern, above all in the context of indigenous
communities, toward what she calls "damage-centered research," lest the
communities in question come to think of themselves "as broken" (409).

Indeed, she argues that "many outsiders benefit from depicting communities as damaged" (412). She writes, "In damage-centered research, one of the major activities is to document pain or loss in an individual, community, or tribe. Though connected to deficit models . . . damage-centered research is distinct in being more socially and historically situated. It looks to historical exploitation, domination, and colonization to explain contemporary brokenness, such as poverty, poor health, and low literacy. Common sense tells us this is a good thing, but the danger in damage-centered research is that it is a pathologizing approach in which the oppression singularly defines a community" (413). But what if all this is a representation of something actually true, of violations and violence? What is the alternative? Uranium-laden wind blows onto Indian reservations in South Dakota. The forests of the Gond and Munda people of Central India are cleared for coal mines. Marshall Islanders suffer the legacy of American nuclear testing and climate change all at once. Rohingya or Yazidis or Roma suffer endless and brutal violence. Favelas are razed, *gecekondus* bulldozed, *jhuggi* slums cleared. In Aleppo you are far more likely to die if you can't leave and have no means of protection because you are poor. This is what is or has actually happened in these places. No novel could be written about such places without accounting for these realities. These are the interstices of structural violence and direct force, the former of which compounds the latter. And in many cases the communities in question narrate *themselves* as in critical situations of exploitation, and there is utility in and cause for and truth in that. How could they not? From a Marxist point of view, not pointing out or emphasizing suffering because it suggests a damaged self might be perilous; my sense is that damage-done-unto must be separated out from the notion of "fundamentally damaged." A move away from "damage-centered research" could itself be crippling, ethically problematic, and even devastating, if pervasive catastrophic subjugation is thereby left unwritten.[3] Indeed to choose not to discuss misery could indicate a certain type of intellectual privilege, embedded in the ability selectively to choose the happy, good, and positive. There are political-economic reasons, real reasons, to emphasize troubled lives and places. Sassen (2014), for example, speaks with conviction of a "warehousing" (56) of the expelled that nonetheless might in the end form new and productive sites for political membership and deserve to be "brought aboveground" and made "conceptually visible" (222).

Furthermore, I am not convinced that ethnography that emphasizes marginality and exploitation suggests that the people in question are fundamentally damaged; indeed, in such ethnographies they are often framed, quite in contrast (and sometimes to a fault, where the discourse slips into heroism and romance), in terms of strength, power, and valor. Think about this question, for example, in the context of street-dwelling children, who do not share a culture, necessarily, but rather a trajectory, a relationship to society, a position of age and space. It makes little sense to wonder at the value of "damage-centered" versus "desire-centered" research in their context (nor, in their case, about "dark" versus "good"). Tuck's prescription, or Kelly's, for that matter, could apply only to whether or not even to turn the ethnographic gaze *to* solo "street children" (or not) to begin with—and it might mandate, if followed, a decision *not* to. The need to assert to these children that they are indeed not damaged, and the presence of narratives of damage (of home, of relationships at home, of family, of self, once on the street), are already presumed, given, indeed already all axioms of the discussion. To remove the notion of "damage" and misery and desperation (and desire) from the narrative landscape would be to fail to heed the complexity of these children's self-formulations and lives. The narrative of damage and darkness should itself be a focus of critical ethnographic documentation and intervention, not avoidance.

Some of Kelly's (2013) critique of the overproduction of ethnographies of "quite how bad life can be" is echoed in Joel Robbins's (2013) call for an "anthropology of the good"—which I am not sure is in fact fundamentally missing from the ethnographic corpus (or even, at that, from "ethnographies of suffering")—to counteract an alleged and potentially limiting obsession with misery. It's a nice idea, and more than just an exercise in rhetoric, except that much of the world's people live in what they themselves define as conditions of misery. Given global political economies, the expanse of what Sassen (2014) calls the "systemic edge" in a system marked by "brutality," much of the world's people do indeed live lives marked by suffering. They are, in Sassen's terms, the objects of global capi-

tal's "savage sorting"—poor, precarious in their exposure to death and illness, exposed to toxins and predation with little recourse to legal protection, exploited in labor, helpless before police or state, forever forced to
move. That is a fact of shared historical conditions. Robbins (2013: 448)
suggests that where previously anthropology took as its object the "primitive" or the "savage," of late

> it has been the suffering subject who has come to occupy its
> spot. The subject living in pain, in poverty, or under conditions
> of violence or oppression now very often stands at the centre of
> anthropological work. I want to trace here the rise of this new
> focus and show how neatly it solves some of the problems that
> anthropologists came to feel marked their work on the other.
> Yet I will also argue that some of the strengths of work focused
> on the other—particularly some of its unique critical capacities
> that were grounded in its grasp of the cultural point—were lost
> in the transition.

Indeed, that suffering subject is certainly at the heart of this book and of the
work that brought it to life; nonetheless, I am not blind to the places and
ways that city-dwelling runaways and their returned or never-departed cognates in the village express and enact happiness, sometimes "in the face of it
all," as it is said, and as Amir said. Walker and Kavedžija (2015) echo Robbins, suggesting that "happiness" gets a bum steer in anthropology, which
"has often gravitated toward more 'negative' forms of human experience,
such as suffering, pain, or poverty" (4). Ortner differs, however. She heeds
Robbins and Kelly and Walker and Kavedžija, despite finding herself initially "startled" by the Happiness Turn, but where they posit an opposition
between the "anthropology of the good" and "dark anthropology," she emphasizes a dialectic, a *relationship,* that might come together in the shape of
a rehewn anthropology of resistance that moves beyond the "misery porn"
critique to something more productive that is about the dark and the good,
yoked together in a complicated world whose precise nature and value do
not need to be proclaimed (or declaimed).

The runaways depicted here certainly embody that—and want to, and say that they would. Among my first questions, as you have seen, is always, Is this better or was it better at home? The answer *always* acknowledges complexity: here, I am free, and by deciding to leave, I (that is, the child in question, in her typified voice) chose freedom and took things into my own hands (*selon* "anthropology of the good"); there, on the other hand, I had not yet betrayed my mother and brothers, I had a home, things were simple. Here, by contrast, I am raped, I face predators and slavery and illness and death (selon "dark anthropology"); there, I was raped and faced labor and illness and death and boredom and traffickers and mines and brick kilns and a desire that could not be fulfilled. So where is better? Neither. Both. Life is difficult, but sometimes it can be beautiful.

Must it be decided? Certainly not. But this answer—a matter the children seem to wish to talk about, and articulate beautifully, as Hecht (1998) emphasizes street children do, to the public's surprise—expresses the need for an anthropology that is committed to suffering and desire and pleasure and aspiration and the various things that constitute *being human* all at once, integratively, and needs not concern itself with taking a position about which angle is better, for it is rooted in the complex realities of lives that are never one thing and need not be so characterized. This complexity, of course—something like the "complex personhood" Gordon (1997) describes in *Ghostly Matters: Haunting and the Sociological Imagination*—corresponds to the complex reality of a world dominated by aggressive markets, fierce exploitation, and profound violence, where most people are *not* rich but where most people experience some good, some optimism, meaning, and hope. Nabil's account of shame for his departure and the loss of his arm, but in the face of the untenable situation at home that kept him from going back, evinces this beautifully. Amir's story, likewise, instantiates a fraught and introspective complexity that exemplifies the need for a synthesis of tales of suffering and hope alike.

Furthermore and following these concerns, the fundamental ethical question is, for us, whether the ethnographer does a violence by telling these stories of suffering, whether they trespass on something that belongs to someone else, appropriating it from them. I certainly hope not. But we can only ask the subjects of the research, enjoin them to be part of the work, and give ourselves to ensuring the work is used to their benefit according to choices, wishes, and concerns they themselves articulate. As Biehl (2005) suggests, in *Vita*, often cited as a foundational work in the anthropology of suffering (and acknowledged as such by Ortner [2016]), "the anthropologist's ethical journey" demands the identification of "some of the ordinary, violent, and inescapable limits of human inclusion and exclusion and learning to think *with* the inarticulate theories held by people . . . concerning both their condition and hope" (20). Again, this representation must start from the voices of those occupying the edge themselves.

Nobody would deny that research that ends in someone not dying or being enslaved is something that the people in question would wish for. The question is whether the moment of fieldwork itself, and the process of writing, brings anything to anyone but the author and some friends who wish to contend or agree with her and a few students. It is unclear whether anyone really wants to be researched. Some people seem happy to be interviewed, to have their lives observed, but they may not *desire* to, actively. It is less unclear that some people want to be written about. People often express satisfaction at having their story told.

And so the answer to the question of whether this work *does* anything for anyone has to be: only if we make it so. And choose to. If that is the thought and the intent and work is done to enact that, and if the aspiration to heroics is an object of critical engagement and is explicitly considered, in a discussion of bias and ethics, then research may have some good to do. Students heading out in the field often fret over this: in such work, it is easy to frame yourself as the hero. You are not. It is easy to frame the subject as a hero. Maybe that is true, maybe it is not, or maybe that is just a question of narratives and perspective. Best, in all cases, to stay away from the epic narrative: if people want their story told, tell it, and do your best to give it (the story) back to them in whatever way you can, or use it to do something they want.

Some meditations in an essay called "The Bard" by Carolyn Nordstrom (2009) do something to address these concerns, give me hope for the notion of ethnography's contribution, give cause for an ethnography of crisis, and help me make sense of this tangle of critiques and defenses of writing darkness—or not—in a rather dark world (see also Behar 1996 for similar sentiments, above all the now-classic assertion that "anthropology that doesn't break your heart just isn't worth doing anymore" [177]). In the face of harsh commentary, Nordstrom asserts that the "why I'm writing" is for her the "kids on dusty broken street curbs . . . who gently and patiently explain what it means to be human, to have dignity . . . whose theories of life are as bright and vibrant as any scholar's I've met. . . . 'We are the story. We are why you travel, why you write'" (35). The counterargument to that initial thought is that, well, it's not just you (the author) at stake here, and the concern is not just "why you write." But Nordstrom goes on to address that by asking, "Do I see myself as some voice, some savior of the war-afflicted? Of the violated and the orphaned? No. This strikes me as offensive. It certainly strikes the war-afflicted as offensive. . . . We all, as humans, have a responsibility to creatively offer something the world" (37). But what justifies, ethically speaking, that responsibility? Actual suffering, of course, and choices we are presented with when we see it face-to-face: not to see it, for example, not to emphasize it, not to write about it. To focus on something else. To focus on "the good"—and in this, it seems, there is a certain vanity, because the choice to focus on the good is more a message to the rest of the academy, or to the world that might read your book, than to the people who are the subjects of research. How would a statement that urges anthropology to focus more on "the good" sound to those experiencing "the dark"? Shouldn't we assume that those in the midst of the latter might worry more about the neglect of their categorical, structural situation—in order to craft a better message about humanity to audiences—than the former? As Nordstrom concludes (in this particular form, as rendered below):

the fact

that the war orphan's story

has seen the light

of day

means

the fight

is

worth it

(2009: 45)

I hold onto this. As you finish this book, wondering whether it is worth it—worth it to read, to write, to dwell upon and think about over and above other things that feel brighter and lighter—I hope you will too.

Most of the people and ideas and places populating these pages are unfortunately more marked by misery, loss, and desperation than pleasure and happiness (by which I mean, of course, not that it was unfortunate that they happened to be the subject of the book, nor that, on setting out, we thought we might discover them to be marked by something else but were disappointed to incidentally and unfortunately find them marked by misery and suffering, but rather that what history has dealt them is unfortunate). I do not wish that, and neither, above all, do they. Wishing that or arbitrary choices of emphasis are not why I chose to write this.[4] The reality of such suffering is about facts of world history and its translation in local lives: most people are poor and precarious and have little power and protection. They do still have pleasures. They manage to do things that might count as the "anthropology of the good." But I refuse to be made or asked to choose which domain this is *really* about, and I know that all these children would too.

And despite this resistance to choosing an affective hue in this way, fundamentally my concern here, in this book, returns to a question posed—or a formation highlighted—again by Farmer (1996, 2001, 2004), alongside others interested in a critical stance on historical suffering, wherein "various large-scale social forces come to be translated into personal distress and disease . . . *embodied* as individual experience" (1996: 261–62) and where "neither culture nor pure individual will is at fault" (2001: 79). Farmer suggests that the sources of seemingly intimate (bodily, domestic, local) harm, adversity, and pain can be distant in both time and place but still kill. For example, a person with AIDS in Haiti, to simplify things, might be dying because of decisions made on slavery and sugar long ago (in the era of empire, for example) and far away (in France, say), and furthermore the death of the person in question is a form of violence, even murder, though it might not seem that way. As with the Haitians whose suffering troubles Farmer, I have striven to ask, What large-scale social and historical forces, from indigo debt to the Green Revolution, from anti-Maoist crackdowns to slavery, might translate *today* to a single person's sleepless tears and torment, to a single home's collapse, to a decision, heart pounding, to leave at night, in the dark and alone, for the train station, for a distant city? This too is structural suffering, not (necessarily) somatic but emotional, familial, and interactive—a psychosocial entailment of structural violence.

And yes, I am in accord with the view that it is problematic for us to make proclamations on the nature of and solutions to Other People's Problems, when we are implicated in the structures of victimization we are critiquing, and when we are imagining ourselves heroes, but my view is that there are some things that are not (or that are less) ethically fraught—that nobody should have toxic effluent dumped into one's backyard, for example, or live in a spill zone without safeguards, or that nobody should be systematically raped by virtue of one's structural position. There is no sense in which one shouldn't call attention to such things and seek ways to stop them by virtue of who one is, because in all cases the subjects do not wish such things upon themselves. As Farmer (2004) puts it, it is not a question of cultural difference, but of structural harm. These are sites where we should intervene anyway, and no matter what. As for me and as for this project, I stand, then (in assessing the ethics of the work, in taking measure of what it means to write this book), with Paolo Pellegrin, who reflects, "When I do my work and I am exposed to the suffering of others—their loss or, at times, their death—I feel I am serving as a witness; that is my role and responsibility to create a record for our collective memory. Part of this, I believe, has to do with notions of accountability. Perhaps it is only in their moment of suffering that these people will be noticed, and nothing erases our excuse of saying one day that we did not know."

Is Running Away Resistance?
Leaving Home as a Historical Act

It is possible not only to be living some aspect of history but to see and understand yourself as embedded in that history, to have the ability to theorize it. These children, despite what everybody thinks about children, absolutely do that. The assertion that runaway children far from home know that some elements of their lives are dark, as in the previous discussion, that they are suffering people, that they have faced misery at home and face it now, again and possibly worse, once in the city, highlights their place as active authors and theorists of their own stories, and thus also highlights their remarkable intentionality in the face of society's view of childhood and demands attention to the defiance they embody and enact of all sorts of structure and form. Formulating a model, a map, a blueprint of one's own situation is precondition and antecedent to a critical consciousness that permits its rejection, its refashioning, and defiance.

Is running away resistance? If so, how, of what, and why?

This meditation is about what kind of resistance children can articulate, despite public conviction in children's inability to resist, about whether and how they can mount a cogent resistance to historical circumstance and subjugation, and whether and how they can act collectively. If they happen to seem to be acting collectively, but without organized coordination, why is that? Can departure itself be a mode of resistance? The effect of such departure is to unhitch the child, or to hitch him in different ways, from the kin networks in which he is embedded at the moment he leaves. What might be made of the rejection of kin structures? One possible answer is that it is not kin structures at all that are being rejected, and that rather it is their disintegration, or the circumstances of their disintegration, to which the runaway's departure is a resistance. Perhaps, later, it turns into a defiance. Once in the city, defiance becomes for these children a way of life, well learned and worn very well. But who is listening? To whom is it legible as defiance? And once the thing being defied is far away, it is no longer defiance of or to the same thing. If the forces, forms, and people it is meant to defy do not see it as defiance, is it still?

Running away is a resistance in the household and often in the village. It is a resistance against local and maybe global norms. Against parents, employers, enslavers, and *potential* employers and slavers, or against *lives* that promise to be exploitive, to enslave, to trap. But is it a resistance against historical mandates, against global orders, against the heritable decrees of structure, capital, and empire? If so, which ones? Given by whom? Formulated how, by the child? Perhaps like this:

My family will never get ahead
I am destined to end up like my dad
I will never be anybody
I will never be respected
    I cannot choose my friends
    I will never get away from who I am in this village
    I will never love whom I wish to love
    my father's drinking is worse and worse
    my father's debt is worse and worse
    my stepfather beats me
        my mother cannot afford my brother's medicines
        my mother will no longer get out of bed after my sister's death
        I've never gone anywhere and never will
        I have to quit school and start work next year
            at the kiln
            at the mine
            at the dump
            at the tea stall
            at the brothel
            in the fields
            in the landlord's fields
            in the landlord's kitchen
            in the factory
            at the quarry
            on the roadside
            (and so on)—
    but wait:
    I have control (I just realized) over my self and my body
        and its location.
    *I **can** leave,*
    *and so: I **will** leave.*

All of these things are directly or indirectly translated from the structural, the historical, and the macrosocial. Such narratives emerge in almost every

runaway's account of departure, and even and moreover in the array of dissimulative and self-protective stories deployed rather fluidly in arrival narratives—and those were themselves revealing of children's awareness even
of the possible matrices or arrays of historical forms that might contribute
to running away, whether derived from their village or their empathic imagining of and actual exposure to other runaways. They are in plain evidence
in the accounts in this book. And they reveal a highly sophisticated engagement with the question of what sort of things might make someone want
to run away, and consequently of what sort of things are out there to resist.
Thus it is quite clear: the (seemingly) simple act of departure can absolutely
be seen as a historical act, in the face of the many forces delineated in chapter 3, from indigo debt to susceptibility to famine to anti-Naxalist purges
to Green Revolution debt to right-wing (BJP, RSS, VHP, Bajrang Dal) manipulations of caste (for example, Dalit/Mahadalit/Denotified Tribe) and
community (for example, Muslim, Adivasi). It is a will to something, an
exertion, unexpected, given countervailing forces. And given the list of
self-aware formulations of departure, as instantiated here, it begins to become clearer that the answer to the vexing question of whether the decision
to depart is formulated by the suddenly mobile child explicitly and consciously *as* historical in some way is unequivocally: yes. Cruel landlords,
lecherous traffickers, abusive teachers or uncles, insufficient funds to save
a parent's life, stresses of rent—all of these figure centrally into stories that
on occasion even take on a Dickensian or Tolstovian character. Granted, the
children, though they might point to the predations of the police, the state,
or certain industries, or to the disadvantage of a certain group or region,
rarely name British economies of jute or cotton, or indentured deportations
to Trinidad, or the construction of the railway as culprit. Perhaps that is
where the scholarly genealogy has something to offer—in bridging the gap
between two scopes of vision, a shorter and longer durée, between the tale
told by the children, historians, of a sort, of the intimate self and the family,
and the historians of the subaltern.

It is clear, then, that departure *is* a form—perhaps the only available—of leverage against historically given structural edicts on who you are
and can be. This is an opting out, a choosing-otherwise—not to be what

history chose for you. And it marks the body and the body's nexus with social space (as in LeFebvre 1974) as a central site of struggle *and* of subjection—just as labor, the body's labor, is taken by Marx (1848, and everywhere) to be the only thing the worker who possesses nothing else has to offer up, in desperation, and revolution her only recourse, once solidarity is found. Both (labor and revolution, or alternatively, for the runaway, disciplinary control/confinement and escape) involve the body as the only real asset of the powerless, but one (in each case) involves centrally its capacity for refusal. In other words, the village child who plans or manages to run away holds nothing to mobilize for resistance except the body, nothing to leverage for autonomy, not even (legitimate) labor, and certainly not citizenship. In fact, that child, given the nature of the half-status of childhood, often barely even has the body to mobilize, as the body is itself often a focus of control, coercion, confinement, sometimes even *the* focus of subjugation that makes the child wish to run away. The child in question is not considered a free and autonomous human being with choice or the ability to exert control over (em)place(ment). The village child's ability to effect intentionality is constrained in a range of ways, as he has access not even to legal status and representation, to freedom of movement, or to money.

Thus, enacting the resistance of running away is exponentially and exceptionally curbed on multiple fronts; this renders it, in some sense, given the push necessary to make it out, even more of a resistance. And thus a child who leaves is always "out of place," because he is not where he is supposed to be: at home, in the village, at school, and so on—he is not even at work, which might itself in another cultural milieu be "out of place." Maintaining that—*staying* vigorously "out of place"—also requires at least as much exertion as leaving in the first place. And that "out of place" is furthermore a shameful marker to elites of a failing society, a society that "cannot take care of its children," should someone see him out on his own. The force of that *where-he-is-supposed-to-be,* because of its universal and decontextualized application, thus also *masks* harm that is being done unto him at home, as it chooses to see only that he is, when home, now in the right place for his age. The response, if he is seen out of place but cannot go back to his family, is to put him in the surrogate family—which is also a hiding place—of a shelter, a nonprofit, or a prison of some sort. That is also one of the places children—unfortunates, of course—are "supposed to be," and also a site of resistance, and indeed of running away.

Resistance as a trope, a central thematic, a paradigmatic and foundational idiom or motif in anthropology has received a lot of attention and a fair amount of critique in the last three decades, in the context both of Foucault's influence, emerging from a Nietzschean tradition, and of a range of others starting from Marxist (sometimes by way of Gramscian and Benjaminian and Wolfian) propositions (see Ortner 1995; Ortner 2016; and Sivaramakrishnan 2005 for reviews of these strands). Much elaboration has been made to such models, accounting, for example, for questions of intentionality (Fegan 1986) or collaboration (White 1986) in the midst of or in contrast to resistance. It is worth considering how we might situate what runaways do in the context of that body (or those bodies) of work, for the corpus is sufficiently well elaborated and broad in its geographic and topical distribution—and it has received and sometimes withstood tests of time and scholarly excoriation—that it can lend a comparative hue to this discussion and help us take stock of the contours, the shape, of human resistance, and thus also of what is being resisted, at this moment of history. On some level, this requires (for better or worse) some work to take a stance—shy of generating a taxonomy, but not of taking stock of how things we observe might fit into structures and relationships—on "what sort of resistance" it is, exactly, that we're seeing. Moreover, it bridges the gap between "dark anthropology" (with its critiques) and the "anthropology of the good" (with its own corresponding critiques).

Though Abu-Lughod (1990) warns legitimately of the dangers of the "romance of resistance," and though many have detracted from the idea of resistance as paradigm, my feeling is that resistance is a good descriptor for something people actually do—and themselves *feel* that they are doing—an ethnographer's key indicator, perhaps, of a concept that can and should be used. People believe that they are pushing against something, mounting grand rebellions or small defiances, enacting friction. They argue, they reject, they make alternatives, and what they do this *against* are real structures of control present in their lives that *we* can see and *they* formulate as targets against which to act counterhegemonically. Sometimes, further-more, they see their *own* resistance as heroic or romantic. Indeed, Abu-Lughod does not, as some others do, plug for a rejection of "resistance" but advocates for a rethinking that emphasizes its potential, following Foucault (1978, 1982), as a beacon that points our gaze to the critical locations where power is exercised. The failure, she argues, of an analysis of resistance is when we see in it or because of it, triumphally, the *defeat* of the structures of tyranny against which it is mounted. Neither, it should be mentioned, are ethnographic engagements with resistance (of which all these reviews and critiques are made) but a fad of bygone days. As Ortner (2016) notes, re-sistance has seen in recent years a "re-emergence" (48) and is a site of such "burgeoning new work" (66) that it can legitimately be said, in the wake of critiques and rejections, that "the anthropology of resistance . . . is back" (62). In these pages, at least, it is too.

In this light, then, how might the resistance performed by runaways be conceptualized, problematized, as a structural response? One way of thinking about children's resistances postcolonially might derive from "subaltern" approaches, as authored, say, by Spivak (1988), Nandy (1983), or Guha (1983) and carried on by many others, from bell hooks to Anu-pama Rao's Subaltern Urbanisms working group. Satadru Sen (2005) and Sarada Balagopalan (2014) deal some with childhood from this perspec-tive. But ultimately few works deal with *this* subaltern *as* itself a subaltern, rather than as a component or surrogate of another subaltern in which they are wards of adults of certain social classes, and fewer still with the agrarian child outside institutional space, a peculiar type of rural subject in his or her relationship to citizenship, subjecthood, mobility, and autonomy. An-other tradition, with which the "subaltern" is connected but not isomor-phic, deals more squarely with the life of the agrarian, in particular as it spans the colonial and postcolonial periods—with the life of peasants, the landless, debt servitude and credit relations, mechanized agriculture and green re/de-volution.

Foundational in anthropological formulations of resistance—and pioneering in making sense of how people engage culture and practice to enact defiance against macrostructural forces in their everyday worlds—

is James Scott's *Weapons of the Weak*, which drew attention to "everyday forms of peasant resistance" (1985: 29) and "hidden transcripts" in the defiance of power. Such "hidden transcripts" are "a critique of power spoken behind the back of the dominant" (1990: xiii) and "expressed safely only offstage" (10), that is, not in public. The dominant or the elite also devise their own hidden transcripts.[5] Such resistance is embedded deeply in quotidian acts of the most personal and mundane sort. It points to the potential to see running away as a practice, a mounted practice, opposed to something. Scott's work, and the statements it makes by virtue of its choice of emphases — and as attested by nearly every review of the development of the concept of resistance — certainly remains the most important and influential statement on the matter.

Are departure and its attendant intimate minutiae — packing, stealing money, studying the train schedule, planning, closing the door quietly, resisting sleep, feeling the night air, leaving a note, leaving no note, confiding quietly in a friend — not also such "ordinary" resistances, in the sense intended by Scott? Certainly they are. That does not at all mean that what is being resisted is itself merely or exclusively present in the dominations of intimate or domestic space, but rather that it is in the historical forces manifest therein. In such a context, the family could be, for example, the locus of labor expectations, of impending years of oppressive work at the family's own hands, at the hands of another family, or at the hands of some force to which the child will be passed on or sold off. The family could also be the nidus of connection to the state — to corrective institutions, to military conscription, to an undesirable relation to certain militias, like the Salwa Judum paramilitary that hunts down Naxalite Maoists. The family could be the locus of socialization to norms and moralities that run counter to the child's own orientations and desires, such as those of sexuality, marriage, or religious obedience or institutionalization.

Certainly, then, these children and the families they come from qualify to be what Scott calls the weak, and the children are often themselves, *within* their families, doubly and again, the weak. Do these, in their multiple nested (domestic, local, national, global) "edges," have any "weapons," such as were identified by Scott in "intact" rural societies? I argue here that mobility itself, that the movement of their sole possession, their bodies, is their weapon, and it reveals a potent resistance that is nonetheless neutered by new forms of subjection in the city (in the context of which, however, the rural—as sign and material reality—as the emblem of hidden origin and the promise of sanctuary or return could itself become again a resistance). As for "hidden transcripts," those too are there, but they are complicated by the fact that so much of what runaways become is by virtue of their emplacement in public. In Scott, agriculture—and the messy complexity of identity in that context—is central. It is here too. Children's departures from the village could be seen as a refusal of, a rejection of, or a resistance to agrarian structures of power that decree village social orders, or that village social orders decree.

The notion of "everyday resistances" points to another domain of defiance, one in which they are in particular likely to occur—struggles over meaning. This, in turn, raises the possibility that what runaways do when they choose to leave and occupy urban space that they are not normatively (by space or age) supposed to inhabit, is contest *meanings,* challenge moralities, engage in semiotic struggles, and that, in turn, highlights their capability to do such a thing, as I describe in chapter 5. It might be possible that we could think, for example, of street culture as a youth culture (but in what sense *a,* in what sense unitary?), in which struggle over the meaning of signs and meanings is central, in the sense that Hebdige proposes, along with Vološinov ([1929] 1973) and others before him—a counterhegemony, a "struggle within signification" (Hebdige 1979: 17). Never mind that Hebdige frets, as Foucault famously and fatalistically did, over the ultimate fate of resistances that are bound to become their own structures of domination, over the eternal cycle of appropriation, repossession, and redefinition of resistance as image (that is, marketable resistance) and resistance movements by society (and in fact, even street childhood can be appropriated and is marketable in various ways); from within that vortex there always emerges new defiance.

I do not fundamentally think that runaway street children form a cohesive "youth culture," but I do think, as I assert rather firmly throughout the book, that runaways engage in contestations over signs, in what

Hebdige, echoing Umberto Eco (1972), calls "semiotic guerilla warfare" (1979: 101), "an assault on the syntax of everyday life" (1979: 105). I do, furthermore, think they are highly conscious of what their actions are "saying" to the world. Comaroff (1985) bears witness to such constellations, in which, by virtue of entrenchment in varying relations of power, meaning is reconfigured and newly charged: "when expressions of dissent," she writes, "are prevented from attaining the level of open discourse, a subtle but systematic breach of authoritative cultural codes might make a statement of protest which, by virtue of being rooted in a shared structural predicament and experience of dispossession, conveys an unambiguous message" (196). Of such resistance, which has the potential to seem to "go 'against nature'" and to "interrupt the process of 'normalization,'" Hebdige further reflects, "The struggle between different . . . meanings within ideology is . . . a struggle for possession of the sign which extends to even the most mundane areas of everyday life. 'Humble objects' can be magically appropriated; 'stolen' by subordinate groups and made to carry 'secret' meanings: meanings which express, in code, a form of resistance to the order that guarantees their continued subordination" (1979: 17–18). If in chapter 5 what such children are seen to be resisting is norms of spatiality, of how a city should look, of where a person of a certain age should be—and should be *seen* or not seen—then here I underscore that, starting even from the kernel of the intention to leave, children are resisting the very notion of who has power over them, even the idea of childhood, the given truth that a child stays home, the given truth elsewhere that a child works, and the idea of home. If home is toxic or violent, says the dominant viewpoint, well, suck it up. That's where children belong. Thus *just leaving* disrupts normative views of childhood itself by challenging the child-family-home bond, by resisting the notion that children should not be alone, by asserting that children can make their own decisions and take their fates into their own hands, even if they are wholly noncitizens, rightless. Hidden inside *just leaving* may also be other resistances, resistances to gendered orders, in particular, but also to class-based moralities or even caste, all of which may be exerted from outside.

In this context, childhood (about whose ontic stability, if we recall

Ariès 1962, we are already rather uncertain) itself becomes the sign subject to contestation, and runaways, by their actions and their comportments and their formulations of the world, deuniversalize childhood, decenter it from its media-bound middle-class prescriptions, in which colonialism plays no small part, and emphasize the local, the particular, the individual in the theorization of the correspondence of place, situation, and stage of life. They simply take "childhood"—their own, indeed, in the form of their visible bodies—and displace it, or rather place it in a *new context,* where it can, in Comaroff's (1985) words, by virtue of being placed in a "novel syntagmatic chain" (197), gain new and charged meanings.

We can thus quite easily establish the presence, it seems, of runaway children exercising, by running away, a certain type of everyday resistance, tacit or otherwise, performed by an individual in a home, against a family or against the strictures of a place or even of labor. We can further establish that it is consciously formulated as resistance in the context of historical circumstances against which the child is at least partially conscious of acting, and even that the child mounts a challenge to meanings and signs through a defiance of regnant norms of where she/he should be at a certain age, and why, and until when. But let me return to another bit of perspective offered by Comaroff: how widely shared is the "shared structural predicament and experience of dispossession" (Comaroff 1985: 196) that children in whom the idea to run away is hatched find themselves in, and what is the nature and ontology of their sharedness? How can we take measure of the degree to which children's resistance might be *collective?* For now I am not plumbing whether it can be organized but only whether it can be collective— indeed, as Abu-Lughod points out, much of the resistance anthropology was interested in, in the wake of the 1960s and 1970s, was "noncollective, or at least nonorganized, resistance" (1990: 41).

It is worth wondering whether the children are aware, *before* arrival in the city, that they are sharing something, a certain mobility and affective experience, with others like them. In the village, how does knowledge circulate about the possibility of running away and its meaning, methods, and currency? In the city, what kinds of new solidarity are born from the circulation of departure stories? Children on the street certainly *band together,* caring for each other altruistically or working in groups, if exploitatively, but they do not seem to mount any kind of political statement on any wider scale; where one can observe "children's councils" or any other kind of decision-making bodies, they seem to be exogenous creations of nongovernmental bodies that are made to look autonomous or independent and intended ultimately to function that way. There is little to suggest a social movement here. But there *is* solidarity: there is indeed a "we" (and a "let's," and other such linguistic indicators of solidarity) that shares a certain type of experience and knows and is aware of the others within its ranks, and there is a kind of web-like, nebulous social domain, a kind of street-dwelling children's dimensional space transecting the city, in which children move, in which they can see things that their experience trains their eyes—but not ours—to see. In this sense, there is indeed a collective resistance against the forces that ceaselessly besiege them.

I often hear people say that street children are (or must be) *resilient,* and that they are exercising remarkable *agency* (see Ahearn 1999, for a critique), and that they enact *practice* (as in Bourdieu 1977). These are surely stand-ins for something else, like the notion that we don't expect to consider as legitimate children's independent decisions to separate from their family, and yet here they are; or that, wow, look at these kids, they seem pretty happy. Indeed, each of these notions—agency, practice, and resilience (along, again, with resistance and so on)—has been soundly critiqued (including, for resilience, by Farmer). And nonetheless, each, even when it fails as a descriptor, probably signifies something real. The question of what it signifies in this case, and the reason it makes interlocutors uneasy, has everything to do with a larger question: can *children* resist? Or rather, perhaps, a change of emphasis: *can* children resist? The basic question is of their essential capacity, their status as human beings able to make cogent sense of the world. They come up lacking in an Age of Reason. They are irrational, incomplete. Even anthropology has not generated a very robust image of children as potentially intentional political actors. Others might make other claims about this, say that the claims here are an overwrought, over-romantic over-read. But the very fact of a child resisting, even just trying, even without departure, is itself a rejection of a norm of what childhood should be, and a profound statement to the world and within scholarship on taking children seriously. And running away is nothing if not that, but rather amplified.

World History and the Runaway Child

The resistance-to-what question—say, for example, that in resisting the family the child is resisting labor, or certain agrarian economies, perhaps even consciously—begs the question of the biggest possible picture, or the biggest picture's connection with the smallest: what or who are child runaways on the stage of world history, of macrosocial forces? What yokes the child to these largest stages and forces, and how exactly does the yoking happen? How is the child drawn into a world system? Of course, the child rides to the city on the channels of labor and labors once there, and the whole landscape joined together by those railways and channels of migration raises the question of just who the "street child" is within the cartography of labor, or outside it. The yoking must itself occur, to repeat myself, through mechanisms like those theorized for structural violence: historical disadvantage at the hands of sovereigns or markets long ago and far away translates to the realities of a day or a life now. Indigo debt to family strife to abuse to child departure. But how can we formulate or conceptualize the place of the runaway child now in systems of labor, production, consumption, and accumulation? Who is she in the cartographic calculus of global capital? What kind of historical reality is the child who departs from home, and how is such a person's life course legible as an element of larger histories? In grand struggles? Are they, by virtue of their young age, bound to be written off the stage and out of history, to be left or rendered invisible while those just older than them write and get written?

Certainly the age-bound aspect of runaways and, more generally, street-dwelling children, and their lack of situational, occupational, or ethnic unity, precludes them from membership in Marx's *lumpenproletariat*, which might otherwise have presented itself as a fine candidate for their classification. It is in the "Eighteenth Brumaire of Louis Napoleon" (1852) that Marx famously describes this underclass uninterested in revolt: "vagabonds, discharged soldiers, discharged jailbirds, escaped galley slaves, swindlers, mountebanks, lazzaroni, pickpockets, tricksters, gamblers, maquereaux, brothel keepers, porters, literati, organ grinders, ragpickers, knife grinders, tinkers, beggars—in short, the whole indefinite, disintegrated mass, thrown hither and thither, which the French call *la bohème*." Marx was, however, not disengaged from the struggles of children, though more often than not the children of his time and purview were rendered vulnerable by virtue largely of the exposure of their parents or families to the vicissitudes of the market and labor's demands. That is in fact true of the children of this book as well, but we consider their trajectory—and their labor—independently of their parents' or the parents' social class. In any case, in the *Communist Manifesto* (1848) Marx writes, "Do you charge us with wanting to stop the exploitation of children by their parents? To this crime we plead guilty." And he decries "children transformed into simple articles of commerce and instruments of labour." In *Capital* (1867), he lingers long on the capitalist exploitation of children's labor, pointing even to forces that underlie what manifests itself now as trafficking. "Now," he writes, "the capitalist buys children" and the parent, in turn, sells them: "he has become a slave dealer." He speaks of capital's ability to create "unnatural estrangement between mother and child, and as a consequence, unnatural starving and poisoning of the children," alongside the widespread "dosing [of] children with opiates" (IV.XV.35–36). Elsewhere, Marx points to running away itself—from labor expectations—by children overworked by the parents to the effect that, "as soon as they are grown up, do not care a farthing, and naturally so, for their parents, and leave them" (IV.XV.119). Tacitly, then, Marx points to a critical nexus between family, child precarity, and market. Gramsci ([1934] 2005) likewise refers to the family as among "the smallest economic units" (281). In Hardt and Negri's *Empire*, family (as a locus of power and subjection from whose disciplines a child might run) is part of "disciplinary society" but is also "in crisis" and "breakdown" (2000: 274).

Wallerstein (2004), most robustly an heir (if critically, in certain ways) of Marx, points to just such a nexus of home (or family) and system, but within a global context, writing that

> Households serve as the primary socializing agencies of the
> world-system. They seek to teach us, and particularly the young,
> knowledge of and respect for the social rules by which we are
> supposed to abide. . . . They have and see themselves as having
> a defined role in the historical social system. . . . Of course, the
> powers that be in a social system always hope that socialization
> results in the acceptance of the very real hierarchies that are the
> product of the system. They also hope that socialization results
> in the internalization of the myths, the rhetoric, and the theo-
> rizing of the system. This does happen in part but never in full.
> Households also socialize members into rebellion, withdrawal,
> and deviance. (37)

Thus, through social-class inculcation encouraged by forces of market and state, households yoke their children to and live historical forces and pre-cepts. But what is illuminated in and by this book is something else besides. A household as a primary unit of a global system may socialize its members either to comply or to resist "the system," whatever that is. That misses the possibility of a third way, where the socialization to the system is the thing that is at issue, and where children may choose to reject it by leaving. Therefore, the household remains a conduit for the global system in equal measure, but here it becomes battleground, flashpoint, for the negotiation of the terms of that socialization, rather than an internally homogenous and generally agreeable node.

Back to it, then. The present case: how can "street children" and run-aways be conceptualized now as actors in and subjects of global systems? What could they be in "the models" in question? One good answer is that they are unequivocally part of what Portes, Castells, and Benton (1989) call the "informal economy" (*not*, write Portes and Castells (1989) in the same volume, "a set of survival activities performed by destitute people on the margins of society" [12] nor solely a domain of the poor). When they left they weren't, necessarily, as they often emerge from impoverished rural agrarian economies, and the *terms* of their departures are usually dictated by those same or other, non-informal economies (like those that pull adult

migrants out of their village, whether flows of domestic labor or demands
for construction work, mining, or any other destination the railway leads
to), but once on the street that is indeed a global economic structure, or
structure type, whose ranks they join. Certainly "street children" do not
themselves *constitute* an informal economy, but rather they are engaged—
and nearly universally so, in occupation, provided they are not trafficked—
in many such.

The informal economy, or perhaps what Sanyal (2007) calls the "need
economy," is most fundamentally marked by the absence of what Portes
and Castells call "institutional regulation." It is in part because runaways
are children, and cannot legally or legitimately be involved in any regulated
market, that they find themselves in such contexts that lie at least partially
outside the licit—though Nordstrom (2004), who also critiques the notion
of the "informal economy," points out that the "licit" and "illicit" (which
she later [2007] synthesizes in the formulation of the "il/legal") co-define
each other in a complex dialectic; the licit indeed requires the illicit. Portes
and Castells emphasize that the informal economy, by virtue of the non-
recognition under the law of its labor, exempts or excuses its producers
from the provision of safeguards that translate to more and better life, that
it licenses or permits exploitation that makes the worker more precarious.
Or makes the worker a child. It can have no wage guarantees. It can "be
undeclared, lacking the social benefits to which it [labor] is entitled . . .
under circumstances that society's norms would not otherwise allow," or
it may allow for "tampering with health conditions, public hygiene, safety
hazards, or the location of activities" in unsafe zones (1989: 13). Portes and
Castells also remark, notably, that "the young often enter the informal market with a definite ideology of individualistic autonomy in relation to the institutions of their parents' generation" with their division into "segmented
labor markets specified by . . . age" (32). Such informal markets, I suspect,
have become increasingly global as barriers to their transnationalization
have fallen away and contexts or mediums for their mobility have increased.
This could potentially mean that a "street child" in Lagos or Bangkok par-
ticipates in something that could be construed as the "same" economy as
her counterpart in Delhi (electronic waste, for example), in part in the sense
that the labor's product, and ultimately the capital it generates, eventually

flows back into a discrete set of producers' coffers, with products flowing into a limited set of markets.[6]

Of particular note for child runaways in the context of the informal economy, and in light of the observation above about the globalization of such markets, is what Portes and Castells term the "unrecorded activities of large corporations"—the invisible unmonitored labor at either end of the productive process, above all in mining and waste. Above all, as I have underscored throughout this book, it is the informal economy of plastic-bottle recycling, the process in which children seek to live (seek livelihoods, that is)—and sometimes end up dying—by virtue of recovering the empty shells of other people's tossed-away, used-up, nonessential commodity and selling them (unregulated) to a "middleman" in some market who gives them whatever price he wishes for the bottles, and by which, eventually, the bottles make their way back to Nestlé or PepsiCo or Coca-Cola. The invisible labor here, which Coca-Cola does not have to know about or say that it knows about and thus does not have to monitor, safeguard, or provide benefits for, thus indirectly profits and, in some sense, is (under)paid by a licit corporate structure and world market of consumption. Thus these children are providing labor for global capital to which the structures of global capital themselves have no obligations and for which they need to do nothing. The emergent and toxic field of electronic waste work is similar—no one at Apple needs to look very carefully at (or even recognize at all) who exactly it is that performs the acid-stripping of copper from disused computers, as it is not officially part of their supply chain or productive process.

Some "informal economies" that such children participate in, notably sex work, trinket sales, shoe shining, or *chaikhana* (tea-stall) labor (for example) are more complicated, more disorganized, or more local, depending on the circumstance. I deal with begging, which can be considered relatively distinct from larger flows of capital and sometimes part of spiritual economies, elsewhere. Amir notes that he worked as a tea-stand laborer, cleaning glasses in exchange at least for food and probably a place to sleep, and then as a domestic housecleaner, an arrangement that also brought him some safety, and he begged. He describes this period, notably as a domestic, as relatively blissful, compared to what came before, but he ran away from it, too, just as he had with a sticky situation at the station:

Recycling life, Jama Masjid

JONAH: So what was, who was the best person you ever met during that time that you were living in Delhi?

AMIR: Yeah, I remembered, man uh, so one time, there was a guy, he worked for army or something I have no idea, he was kind of like a general, what do you call it, a general?

JONAH: Yeah, yeah.

AMIR: So "general." So one time on the street I brought him tea, so he sort of started liking me and he just asked about the family. So he put me, he said do you want to work in home, so he put me in a home, to work for the family, to do cleaning, cleaning like you know do laundry, things that were a pain for them. I think that was the best time I ever had because you know they, he treated me really well, he didn't have a son but he had two daughters, they were all my age, so we were just like having great time.

JONAH: So you were kind of part of his family.

AMIR: Yeah, so that was the best moment, but at that time, my name wasn't Amir at all my name was Hindu name my name was Gautam.

JONAH: So your past, it wasn't known to everybody.

AMIR: No yeah I had couple of names changed while living on the street.

JONAH: Yeah I think that's very common right?

AMIR: Yes, so, they didn't change my name but I change myself my name because I knew that like if I was going to work in a Hindu family and they won't want to have a Muslim kid, anyways.

This of course forcefully recalls the discussion above of dissimulation, naming, and communal or caste identity—and its linkage to attempts to erase or blur the past.

JONAH: Yeah, yeah. That's interesting. Did they ask you about where you came from?

AMIR: Yeah you know they ask me about it but I didn't tell this person where I came from, just like told them yeah I ran away you know, my parents are poor here, they just didn't really care, because I was just working with them, they kind of got a servant in their home who can clean, who can just like, you know, clean the floor, who can just do the dishes, who can just go out, bring the milk, who can do stuff. It's not like costing them anything. Only costing them like they are providing me food and place to live.

JONAH: And that was a nice time for you because you were safe.

AMIR: Oh man, that was like, also those two girls, like they would play, and captain would let me. Like sometimes he would buy things for me, or like sometimes if it, buy things for me, like his daughter would buy things for me, like tell him how he [Amir] didn't get it so can you buy this thing for him too? So they were super-nice, like they had to treat me equally.

JONAH: Wow that's incredible. So why did that end?

AMIR: So . . . why did that end . . . I don't know why, one day, I felt that, I felt that you know, ok let me remember why. I just felt like I wasn't really belonging to them, sometimes I felt, I didn't really feel happy, and just like one day I decided I was going to run away from here, just like you know, they were really nice people, like they didn't really treat me bad at all, just somehow, after few months I didn't like living there anymore. I felt alienated.

A series of nested runnings-away, in some cases driven by peril, and in some protection. Later he even fell in, by his report, with a drug smuggler and headed for the Bangladesh border, which he fortunately was blocked from crossing.

He had some sort of like drugs for money, and he thought I was a small kid so I could help him pass the border. The guy, you know, he got caught and you don't know, the bag I had, the bag had enormous amount of drugs, enormous amount of money. I, as a kid, had no idea, I was naïve, I had no idea, no idea . . . he trusted me, he thought that you would be, you are the perfect, I mean I didn't really know that the bag had, certain kind of drugs. Because he said, like you can help me with this, I'm going to give you a lot of money and he already gave me a lot of money and then he was also taking care of me very well, like you know he would just, on station or anywhere, he was just taking care of me.

In this way Amir briefly, by virtue of his vulnerability and helplessness and need for protection (clearly also a motivator when he cleaned house), could have been considered a pawn in one of Nordstrom's (2004) shadow economies.

Some children find themselves moving through a chain of many informal and often exploitative informal labor situations — sometimes moving in and out of other spaces providing a sense of protection, especially the sorts of religious sites I discuss in the previous chapter. Consider Anjeel from Roorkee (Khushboo is denoted as "K"):

ANJEEL: I ran away from home and got to Amritsar. I started working and meanwhile also learned Punjabi. I was working at a shop. One day I ran away from the shop because I got into a fight and the owner beat me with a stick. When I left, he [the owner] came to the railway station to get me back but I left Amritsar. I used to live in the owner's house. He gave me Rs. 900 per month and I used to work 10 am to 5 pm at his shop. After work I used to play with my friends who lived in that area with their families.

K: How did you meet the owner?

ANJEEL: He found me at the railway station and asked if I would like to work at his shop. He took good care of me.

K: How long were you in Amritsar before the shop owner found you?

ANJEEL: I was in Amritsar for three to four days. I used to go to Golden Temple and eat there. I would do "seva" [service] in the Golden Temple and would sleep there at night.

K: What did you do after you left Amritsar?

ANJEEL: I worked for four months at Amritsar and after that took the train to Haridwar. I stayed there for two days. After that I came to Delhi.

K: How old were you when you ran away?

ANJEEL: I was nine years old and now I am eleven years old.

K: When you left Haridwar, you came to Delhi, then what happened?

ANJEEL: I came to New Delhi from Old Delhi, I was hungry and looking for food. Then I found a big guy called Abdul and I started working/staying with him. One day I had a fight with Abdul because he wouldn't give me money. I worked with him (made and sold water bottles), he fed me food but didn't give me my fair share. Then I left him, and started collecting my own bottles. I would earn Rs. 100–150 every day. After I separated from Abdul, I got into drinking "solution" [solvent]. One day I was high on solution, and fell asleep on the railway track. A train came and my little finger was severed. There was a lot of blood coming out. Then my friend took me to the police station where Raju, who takes care of us, took me to the hospital and in the morning I got surgery done. After that I returned to Chaar line, where Salaam Baalak Trust found me. When asked, I was reluctant to go home because I thought my parents would hate me and send me away. But [NGO worker X] convinced me this would not happen and now I am going home.

K: Then why didn't you think about going home?

ANJEEL: I was scared to go home.

K: Who were your friends in the platform?

ANJEEL: I had many friends who had also run away from their homes. There's this kid who's about eighteen to twenty years old now, he had run away when he was less than nine years old. He thinks now if he goes home, his family won't recognize him.

Stories like Anjeel's illuminate the gray area between well-organized exploitation consisting of large-scale networks of human traffic (with relatively centralized flows of labor, profits, and commodities), and small-scale, opportunistic, and scattershot engagements (with vulnerable children as relatively expendable labor assets who are subject to physical and emotional discipline and captivity, and who may well flow into trafficking networks proper). Eventual exit from such subjugation, if and when it ever happens, is often accompanied by trauma and shame so strong that the person in question never goes home; that might require a too-painful reckoning with the chasm between one's past and present.

Children who run away to find themselves sold or stolen or trafficked into slave markets for sex, manual labor, agricultural labor, or domestic work are indeed often part of world economies, and their situations are tied to or the same as those of runaways, if in part beyond the purview of this book. Nonetheless, the boundary between "running away" and "trafficking" is very murky—poor village children can be enslaved, but runaways are also susceptible to traffickers in the city, by virtue of being alone. Children told unending stories of being trafficked into such situations both directly from home and after having already run away: through tales of child dealers operating sometimes in the countryside; sometimes in the city, waiting for new arrivals; and sometimes in between, in depots and in transit. Sometimes trafficking can appear informal, even in the domain of the domestic and familiar, and thus disguise itself as benign. Children may be stolen or their parents may be tricked or convinced into renting them out or selling them. Children also told tales of murky people who "traveled with them" or "brought them" to the city. Roshan, from the village of Siwayapatti in Muzaffarpur, Bihar, recounts, "I came with someone who I met after I ran away from home. That man was going to Delhi and offered me to go along with him." It is probably impossible to tell what happened here, so we have to take such accounts as they are or perhaps what we suspect them to be. Ashok's tale belies the complex relationship between parental involvement in trafficking, labor, and violence. Ashok, who is from Motihari, also in Bihar, explains this in a less ambiguous tale: "My father sent me to work," he recounts. "So I came along with someone from my village." He soon adds that "the person who brought me here [to Delhi] used to beat me with a cable wire. . . . He got me working in a factory, where I worked making ladies' purses, attaching the string to the purse. There were many kids with me. One mistake and they used to beat us really hard. So when I got the moment I ran away." And listen to Amir, again, as he speaks about such forms:

JONAH: Did you ever know anybody who was trafficked or sold, almost as a slave?

AMIR: I think so, I think so. I knew a few people I don't know if they are still there or not. . . . Yeah, yeah one kid, his kidney was taken out. One time, he was, you know, he said some people came to me and approached him, they took him to sugar mill and they told him he will work for the sugar mill and later on he was back on the streets and like his organ was gone, was taken out.

JONAH: Were you afraid of that kind of thing? How many times a week do you think, or let's say a month, while you were living in the station, would someone try to get something from you, whether it was labor, or going to Bangladesh, or kidnapping you, something that like you had to put up your defenses, like in a month how many times?

AMIR: Like, you know, every day I would defend myself. Like it was a battle, living on the streets was a battle. You were like fighting for your life every day. But like traffickers were like once or twice, in a month.

Economies of child slavery and human traffic are tied to the larger world markets also faced by runaways by virtue of the fact that similar or the same children are susceptible to both, and that the children's *position* of susceptibility is in part created by those markets, but also that the markets *into which* they are sold, for example at coal mines or cotton fields, are themselves global. It is worth saying also that street-dwelling children in India are often *consumers* within "informal economies," for instance of narcotic solvents (glue, white-out, paint thinner) or other narcotics. In some places, such children might also be entered forcibly into markets of war labor or other forms of violence. These, too, obliquely, serve as channels by which dislocated children participate in world-historical systems.

Ferguson (2015) takes issue with the models, as they've been applied, of both "informal economy" and "lumpenproletariat." The vigorous critique he mounts in the context of economies of distribution in southern Africa focuses on presuppositions by which Marx framed the latter (the lumpenproletariat) as a kind of inchoate "proletariat-in-waiting . . . or else a kind of social refuse, of neither economic nor political value" (92), of little interest in its own right as a distinctive historical form. As for "informal economy," he points out that the term has become so vague as to be rendered useless and to refer to nothing in particular. Ferguson is interested in a nexus of "precariousness and flexibility" that might rather be characterized as "improvisation under conditions of adversity" (94), with an added layer of "survivalist" strategy, a fine descriptor for what child runaways can potentially do, age constraints notwithstanding; if limited in value, it does something to begin to characterize where such people as runaways might fit in schemata of global capital and their larger historical landscapes—in Braudel's longue durée—where before, perhaps by virtue of their exteriority to legitimate flows of production, they received precious little attention.

But most fundamentally, what child runaways really are is *at the edge*. At the edge of the rural and the urban alike, the world-historical landscape and the local, the economic and the spatial. Indeed, they are at the edge ideologically, of what anybody *thinks* of as childhood, of conceptualizations of youth and its institutional and normative definitions. They are at

the edge by age, by place of origin, by their outsiderhood in the city, by
their renunciation of family, by their nonpossession or illegitimate possession of capital or the means to earn it, by their potential offering only as a
commodity to exploit. There is perhaps no better descriptor than this for
them. But if we posit the existence of a social edge, we presuppose a certain unity: that it is indeed *a* thing, that it has some internal cohesion, that
people occupying it see something like what we have described as what
they occupy, and that various groups see each other as commonly occupying the limen. In general, I believe the evidence offered in this book sup-
ports the idea of a cohesive social edge that recognizes itself internally in
that way; the children here certainly know which world they are part of and
which they are not and never can be, and they know with whom they share
that zone.

A runaway's place (or the place of a mass of runaways) in world capitalism—and, more generally, world histories—is tremendously hard to
characterize; indeed, such subjects as runaway children living solo, unhitched from kin, in distant cities far from their rural homes, provide a
general problem for models of production in which everybody's place has
some meaningful relationship to a system in no small part because of just
how profoundly marginal they are, how entrenched at the edge of things,
and further because of their consistency (consistent process, consistent
presence, consistent dynamics and experience) and their numbers (for they
are certainly not a rarity, though they are certainly not all from the same or
even necessarily similar places). On a most basic level, however, it is reasonable to say that they must be part of, or emerge out of, what Tania Li (2010)
describes as the "dispossessory dynamics of agrarian capitalism" (398).

In particular, it is not the grand dislocations of global capital but the
"piecemeal dispossession of small-scale farmers" (385) with which Li is
concerned, in the context of a capitalism that she sees as rather fragmentary. Histories of the sort of edge that runaways form and from which they
begin are marked by genealogies of manufactured debt enacted through
indenture, agricultural peonage in land tenure, and other forms of near-
enslavement. Following Breman (1983), Li engages with the prevalence of
a "rent capitalism": "when taxes or rents are high," she writes, "when sub-

sistence crops fail, or when the price fetched by cash crops does not match the cost of food and inputs, farmers are compelled to borrow money at high interest or mortgage their property as they enter a downward spiral" (2010: 386) Such a process is precisely what I assert forms the antecedent to the family stresses that make it difficult for a child to want to be home with family. There is no doubt, however fragmentary capital might be, however much it might be considered "not a singular form or force" (Li 2010: 400), that debt, as Graeber (2011) suggests, constitutes a ferocious, long-standing, and relatively regular element of the global experience of rural people, in particular in their nexus with the urban, and a strain that requires very little to be converted or translated into an intimate affect, emotion, and personal desperation that might pervade a home.

It is Sassen (2014), however, who most carefully and explicitly unpacks the notion of a global "systemic edge" born largely of markets, and their incarnation in the political, whose core and invariant feature is "expulsion" (211). A periphery so envisioned, she writes, is crucially *not* the one we are accustomed to—of territorial borderlands of sovereign states. It is indeed not precisely or strictly spatial in that way, even though it is necessarily always spatialized. Rather it is an economic edge born of "the move from Keynesianism to the global, era of privatizations, deregulation, open borders for some, [that] entailed a switch from dynamics that brought people in to dynamics that pushed people out" (211). The notion that emerges is that the systemic edge produces Sassen's "savage sorting." Sassen underscores the translational elements of market processes that go beyond what might be rendered as statistical growth or decline: children abandoned by desperate parents, for example, or suicide (214), the very type of translation discernible in children's departures from misery in the village to subsequent misery in the city.

I am taken likewise with the great Bengali economist Kalyan Sanyal's (2007) construct of the "wasteland," a concept that speaks quite well to Sassen's "edge" and Li's "dispossessory dynamics," both as complement and contrast. For Sanyal, there is an exteriority to capitalism that neither opposes nor attenuates its mass or force but that also "challenges [its] universality." Indeed, the denizens of the wasteland, unlike the Marxian "surplus army," do not fit the models that guided the authoring of notions of capitalist relations of production, even if their existence is a squarely Marxist fact; they "resist being captured" in any such formulaic calculus, as "their exclusion from capital is permanent" (55). Historically, the empty spaces to which lepers were relegated were at length filled with "the new incarnation of the lepers," constituted of "the poor, vagabonds, criminals" (44); this is where, in the postcolonial context, "the dispossessed and marginalized" are made to "wander around in a wasteland created by 'capitalist development'" (45) that is "inhabited by the excluded" (47).

To Sanyal, it is not only that the likes of child runaways and other bottom-of-the-heap outcast(e)s are exterior *to* capitalism, and not only that they are exterior to it (of course) *because* of it, but rather that it is, itself, essential in the development *of* capitalism. The productive capacity of the wasteland's inhabitants has been decimated by capital's incursion, and they are granted no passage back to the inside. In this vastation, the "rejected, the marginal, the leftovers of capital's arising, the wreckage and debris," as Sanyal puts it, "remains as an outside" of the system of capital (53), but also in a dialectic, co-defining relationship.

Sanyal affirms the supposition that there are some so external to processes of capital—and whose work is so anomalous in models of production—that they cannot be well classified by existing models. Hence his view that the wasteland is so outside that it is beyond, and unrecognized by, even the basic constellation of class relations (2007: 58). His contribution—no small fact, for this account, that it was crafted from Calcutta—is the notion that despite this exteriority, the development of capital relies on the wasteland in part by virtue of its stalwart maintenance of what he calls "non-capitalist production" (59) and in part by virtue of the fact that such a space of exteriority countermands claims for the universality of capital as model. "Capital," he writes, "can then posit itself as the universal only by denying the existence of the outside: an act of discursive violence" (62). In other words, for our purposes, capital cannot accommodate the beggar, the trinket vendor, the prostitute, or the urchin, even while capital needs all these to become what it becomes. And thus there emerges the possibility of something else beyond capital that is nonetheless in eternal dialectic with it—something that is not even of wage-earning, citizenship-bearing age and

Wastelands/edges: (*top*) Yamuna River silt islands, Delhi; (*bottom*) dump/slum, Sarai Kale Khan

status that collects plastic bottles and begs and steals for a living, indeed something, as I have intimated, that is not even "given the option to perform any labor for capital" (63).

What, then, does the leper, the vagabond, or the urchin do for the development of capital? What do they offer? Very little, from the vantage of conventional models focused on labor's interaction with systems of production—only in the prior generation (the parents) or the later iteration (the former runaway as adult) is that at stake. The child whose contribution to labor is in his capacity as plastic-bottle collector, for example, can only cycle elements jettisoned from capital's machinations back in so that they can re-enter labor proper, but at that stage there is little to no contact with the producer, nor with the consumer, nor even with the labor involved in manufacture. And so indeed the likes of street children do not fit, they are outside both theories of capital and capital itself. And yet there still is something: something at once more complex and more subtle in the relation of such extreme marginality, along multiple axes, to capital. In part, it is that they form an outside in contrast to which something and someone can be *inside* to begin with. Capital pushes them to that outside but also keeps them out by virtue of their new position, and despite the historical specificity of their circumstances of dispossession, agrarian and otherwise, from setting to setting, once in their new habitats the formation called street children looks remarkably similar across contexts worldwide, and thus it is indeed part of a *global* systemic edge that manufactures similar outcomes if not situations of exclusion. There may be some solidarity, in part for this reason, *within* contexts at this edge—between street children and between street children and others with whom they share space—but even with the cyclonic force of various media there is and perhaps can be little to none between street children across contexts, except when actively produced or performed by international organizations of aid, advocacy, and child rescue (such as it is, so-called).

And so all the while street-dwelling children's exteriority, like that of others on the wasteland, is critical to the making, the formation, of the system. But unlike some of those others, for example the generational vulnerability of the Adivasis who inhabit (or, in some cases, who inhabited) India's mine-ravaged forests or the *mahadalits* (the ultra-untouchables), child runaways and certain of their compatriots of the wasteland (migrant ayahs, disabled beggars, transsexual performers) are not only *dis*enfranchised and *dis*possessed but (once they are legible as incarnations of a certain form, that is, as street children) never enfranchised and always external to begin with, from the minute they inhabit that identity, even when what they *come from* is *historically* and more enduringly a situation of dispossession and when the fact that they are there has much to do with that. There is thus more than just structural disadvantage or systemic exclusion here—rather there is structural and then a situational disadvantage *combined* to make a doubly potent sort of exclusion, a perfect storm for exploitation. I want to keep those two marginalities conceptually separate until they converge and no longer can be. The combined result is an anomalous position that stems from an age-bound status and inherited dispossession (that itself may in fact be subsequently heritable, in the case of legacy street children who themselves have children on the street).

Nowhere to Run: Leaving Home, Longues Durées, and the Life of Indeterminacy

If Marx and his interlocutors—some of them incarnate in the ideas unpacked here—have made one thing clear, in the age of ideas they launched, it is that history, following Farmer and many others, begets life and death, that the movement of empires and the whims of viceroys can find their way to the intimate world of a twelve-year-old in a collapsed village of Bihar or Bengal, can dispatch themselves thence to a child's experience of body and affect. History imprints itself on lives; it may shape, but nonetheless it does not determine, for even such as children (subjects, not citizens) read history with their own eyes and interpret it, and they have their own replies somatic and emotive, their own response, their own defiance, and if the body, its movement, and language are the only resources they have at their disposal then departure by its very nature is the mounting or articulation of a small rebellion, a response to the way things are, a statement that this is not how things have to be, nor *where*.

And so it is that the child leaves and so it is that the leaving—and the insistent being there, where she ends up—constitutes a small revolu-

tion and perhaps begets more such. The child has made a big and difficult and unlikely choice and has refused and defied the forms of power
against which she has few tools of resistance. That took a tremendous effort
and will against something holding tremendous force—institutional, cultural, political, economic, and historical. It was a remarkable defiance that
evinced a child's ability to take her fate into her own hands, a resistance by
someone with nothing at all but her own self against multiple nested forms
of domination.

That sounds good, very nice as far as it goes.
But on the other hand. And yet.
In both places, home and the city, lie traps, or both places *are* traps,
and this reality evinces and underscores the sad situation of the child's subjection, confinement, choicelessness: everywhere too, wherever they turn,
is unfreedom, oppression, and the child is no citizen and has no rights
and indeed sometimes, when trafficked and enslaved, has still less, has the
opposite of rights, has only others' rights exerted on him, being owned as
he is; in the end, perhaps, in such a world, there is nowhere to run. Wherever you turn, they're waiting, lurking, out to get you. Damned if you do,
damned if you don't, nowhere to run, nowhere to turn, nowhere to hide,
checkmate, cornered: the child's bane (any child's, but especially these),
always subject to and subject of, never entitled to or endowed with.

But still, let's check back in after a decade. Time passes, things
change, and even the world sometimes changes. Children become adults.
Children resist the new power they face by running away *home,* where
sometimes they command new respect. Sometimes they keep another kid
from leaving. Sometimes they live, sometimes they save a life. And things
might turn out better than you think. I'll be looking for them, those who
lived, and tracing their path as best I can.

So it is, rather than over, to be continued.

The solo child's place in history is an edge, indeed, and indeed perhaps maximally so—an edge of households, of regions, of global econo-

Makeshift HIV/AIDS clinic, Kashmere Gate ISBT long-distance bus terminal. Health workers showed me a register logging visits and results from many children who suspected they'd been exposed. Nearly all tested positive.

mies, of train stations and shrines, of other children's spheres, of everything, really. And not just of these spaces and domains, but also by virtue of who they are in any given moment: the child runaway in the city is at an edge in age, in situation, in material assets and kin connections, once far from home. Indeed, at the edge of life itself, at the edge of everything. They are precarious. Perhaps that is indeed why they run and, once they run, if they are unlucky, why they die or decline at the solvent's mercy, pain and a former self hidden behind a rag. For what do they have to turn to, what recourse, once in that snare, that goblin's lair? Was it a trick, the promise of the road, of something better, of leaving? It was a trick, yes. Many children come to feel that. It was a ruse, that idea of freedom, that image. What else to expect for such children at history's and capital's edge, or beyond it?

Not everyone is so terribly unlucky, of course, as all that, but to not be, a child needs some other path back to the inside, a pathway that is looped into networks of wealth, capital, normalcy, and enfranchisement.

And that requires the donor-fed NGO or the state-enforced spaces of policing incarnate as confinement and discipline — the reformatory's heir. Labor
there (at home), labor here (on the streets), rape there, rape here, beating,
exclusion, illness, sadness: everywhere.

But running away, as I have said over and over, is complex, like anything in life. People are complex. It is not — it cannot be — all one thing. Its
outcomes, moreover, cannot be divined. In one interview, Amir, looking
back and taking stock, was tentative:

> JONAH: So what would you do — let me give you a hypo-
> thetical — what would you do if you went home next week and
> a fourteen-year-old, like you were, comes up to you and says,
> "I want to run away, can you tell me what it was like, was it
> great?," what would you say?
>
> AMIR: I would say no.
>
> JONAH: You would say "stay in the village"?
>
> AMIR: What would I say to the other kids that want to run
> away?
>
> JONAH: Yeah.
>
> AMIR: No, I would say no, because life is like, I would say no,
> I just got lucky, in many situations I could have been dead too,
> and I'm just being honest, because I got into drugs, I got into
> stabbing, I got into a lot of different crazy things, it was just like
> you know, my life was kind of that way that somehow I was able
> to pull out through, pull out through all conditions.

Note that he says *through* all conditions — a triumph of the individual will,
as he narrates it, and as it may in fact need to be, over the structure and circumstance that is itself produced in part by what at least seems like an act
of will over structure and circumstance.

And lest things get too pessimistic, it is also possible on the whole
to accommodate the view that running away is not *fated* to end in misery;
it does not need to be judged a mistake, or indeed to constitute a mistake.

Would we have asked Huck to stay with Pap, after all? Amir elsewhere provided a glimmer of hope that misery need not pervade the runaway's entire tale of an escape from seemingly inevitable doom; indeed, this same person who said he'd never endorse a child's decision to leave offered a sign, if simple, that complex personhood and the poignant depth of being human in a matrix of other humans, and of history, is what it's really all about:

> JONAH: Do you regret having run away? Or do you feel like it has made you who you are now?

> AMIR: I mean I think that I don't regret it at all, I think that was the best thing that ever happened in my life.

But Nabil's tone on the question of regret, of turning back, of taking it back, and of how to take stock of things, was rather different from Amir's, and perhaps more representative of most children's experiences. This is the Nabil whose voice you hear in the first pages. Once Nabil is in Delhi and loses his arm to a train, then it is the fact of running away itself that becomes the ground zero of crisis, the things-having-fallen-apart that prevents him in shame from return. So when after weeks gaining Nabil's trust I finally record and interview him by the Yamuna banks, asking him, Was it a mistake to come here or not?, he does not wait for the end of my question to utter no.

No. If I'd stayed home, he says, I would have kept my arm.

But other than that, I now say, was it a good idea or not?

Not.

What is life here like?

Worthless, he answers.[7]

Taking stock, Amir says one thing, and then says another. Nabil, for his part, says something altogether different. And throughout the book, such ambivalence and uncertainty appear again and again. Stepping back from such stories, and such assessments, one is perhaps left uncertain.

What, then, to think of or take from all this? One should, after all, be able to close a book and feel a nice fable, with a nice moral, has been told, and then smile or cry accordingly.

Imagine a map of these children's aggregate movements; imagine it animated, like a moving image of ships at sea, or luminous flights aloft: ten thousand similar responses to shared conditions, flowing along common currents of human movement, each of which seems and is its own distinct journey, a long and indeterminate limbo between home and the city, conducted by the railway and drawn by history, between childhood and whatever comes next. Sometimes, but not all the time, they flow back where they started, but in that direction the threads are less dense.

There is no simple end to this story, no grand foe to overcome, no clear-cut happy conclusion to tie everything into an easily understood bundle, no easy, pleasing heroism. Not even a well-resolved Right and Wrong to give it a neat shape. One is left with what amounts to a chronicle of questions, an anthology of unresolved stories. It is indeed far from clear that these tales are driven by some single "cause" that can be identified, or countered by any single strategy that would "solve" their miseries. We'd be led astray to latch onto any such thing; the runaways themselves would be the first to tell you that. Indeed the story, or the corpus of stories, is not ours to lend shape to; it is not to us nor to anyone else to say what should be made of all this, to decree what it means. It is to the *children* each to give it a distinct arc, both through their actual experiences and, at different moments relative to their journeys, both on the road and much later in life, through their secondary accounting and interpretation of such movements. At these different points of reckoning, the runaways or former runaways may tell it differently, may tally or weigh the experience in variable ways. They may, even, at each of these points, themselves not be sure what should be made of what befell them.

And in the midst of this indeterminacy, of all this ambiguity, something nonetheless comes through, a glimpse of something fundamental comes into resolution: a troubled landscape of history and of suffering, a landscape not only that intimates or conveys some meaning when viewed both from afar and up close, but that is *given* meaning by the children themselves, by the movements they themselves set in motion. They do not merely reflect realities; they create them. This much can be said with certainty. Even if that creation is an act of desperation, and even if it produces more such acts. The movement, the running away, is an inscription on a historical and cultural canvas, on the map of time and place itself. By such movements, the children are not only the subjects of stories, but their authors.

Chapter 1. Nabil's Predicament

1. It is also, notably, home to the curious new "five-star" Pakshi Resort, "a complete leisure zone that blends corporate flavor into a holiday environment" where "you can truly enjoy all elements that sustains the life on earth; sunlight, fresh air, good foods, sky, the clouds, the flow of river, the seasons, or to enjoy misty morning, or to get the feeling of fogs, or glittering moon, twinkling stars."

2. Which is, of course, not Nabil. I use pseudonyms for all subjects unless their stories are by some other means publicly accessible, or unless they are adults acting in an official or otherwise publicly visible capacity.

3. Nabil is not the only child whose return I have traced; in a later chapter I will discuss the runaway Nuruddin, from Cooch Behar, West Bengal, who disappeared one day in 2007 and whom I found again three years later, much changed, and with his family in tow.

4. It is notable that staff from at least three Delhi NGOs attributed running away to conditions in "flood-prone areas."

5. I did not know then, of course, what I know now from my Roma-centered research about the ways that Romani "street children" are usually situated and well embedded in family groups. My newest NSF-funded research is on Roma as objects of segregation on Marseille's polished streets.

6. The Jat are a poor and often landless low-status ethnic group of the Sindh and Panjab.

7. Properly speaking, this is one of his character's assertions. And I modified this, past tense to present. I originally saw this emblazoned in polished metal, and in French, on the floor of the Montréal Musée des Beaux-Arts, as part of Betty Goodwin's permanent installation piece *Tryptique: Chaque question possède une force que la réponse ne contient plus.*

Chapter 2. Home and the World

1. In this case, a didi is either an auntie-like figure who acts as a caretaker or a social worker (or NGO representative) who takes on the child.

2. The interview is in Hindi, but he uses the English word *death* here—as a Hindi lexical item, in predicate adjectival form.

3. The film *Siddharth* by Richie Mehta shows how such trafficking "at the hands of one's family" may itself be accidental, when the family is unsuspecting and vulnerable, and thus easily duped.

4. It is interesting in this sense to think of the renamed street child truly as a new incarnation, a newly fashioned self: the term *avātara* indeed is derived from the root *tr/tar,* which refers to crossing (downward, as in descent); these children have crossed not a boundary between physical lives (though many have come close) but rather a boundary between conceptual lives, a feature of which is certainly a true starting over.

5. *Baba* refers to a range of village healers, shamans, exorcists, spirit mediums, medicine people, and so on who are consulted for help in the face of various predicaments and crises, and sometimes for mental or even physical health problems.

6. *Harām ka beta,* a bad kid, bad seed. *Harām* refers to that which is forbidden or nonkosher in Islam.

7. *Halāl ka beta,* a good kid, an upright kid. *Halāl* denotes that which is approved and moral in Islam, akin to "kosher."

8. *Bihari* refers to someone from the impoverished state of Bihar, but in this village context, in Haryana, it also will denote a low-class status, membership in a family of outcast(e) pariahs, probably menial or domestic laborers of even more base status than village residents who are laborers from that region.

Chapter 3. Of Crisis, Calamity, and Village Fabrics

1. The film is the product, notably, of a theater workshop for street children Nair was involved in. *Salaam* won the Cannes Camera d'Or in 1988 and was nominated for an Oscar. Most actors were "real" street children. Sadly, the lead in the role of Krishna, Shafiq Syed, failed to secure further roles or jobs in cinema and attempted multiple suicides in the years following the film. He now works as a rickshaw driver in Bangalore, where his decline makes him the object of occasional ridicule but also, for example in the wake of *Slumdog,* occasional media attention. The film's profits helped establish the major organization the Salaam Baalak Trust.

2. As in, say, the *Manusmriti* (56, *Bhagavata Purana,* xi. 5–11): *Kapala-patra-nirata / Kangkala-malya-dharini.* The "garland of bones" is associated with Shiva, who in the text is described as "wearing a Brāhmanical thread composed of white snakes, clad in an elephant's hide, with a necklace of beads, and a *garland of skulls,* riding upon Nandi, accompanied by ghosts, goblins,

spectres, witches, imps, sprites and evil spirits" and elsewhere, in a smear by Daksha, as follows: "He roams about in dreadful cemeteries, attended by hosts of ghosts and sprites, like a madman, naked, with dishevelled hair, wearing a *garland of dead men's [skulls] and ornaments of human bones*" (as in Wilkins 1882).

3. Brunner further writes, "The Persian word āwāra (Tehran: āvāre) is certainly common in Iran. It is already in early New Persian, in the Shahnama of Ferdowsi ('lost, in flight, vagabond,' Fritz Wolff, Glossary . . . , pp. 35–36). Both *āwār* and *āwāra* occur in medieval literature (examples in Dekkhoda, Loghatnāma, fascicle Ā—Abu Saʻd, Tehran, 1946, pp. 196–97). I don't recall it in Pahlavi or Manichean Middle Persian, and it is not in Durkin-Meisterernst's dictionary. Although I don't have an etymology, for the morphology, cf. Mid/New Pers. āwāz 'voice,' āwām 'time period,' Pahl. and Jewish Persian āwādag 'generation (according to Mackenzie, Dict. of Pahlavi, p. 13). Confusingly (or, for a poet, pleasingly) similar is *wīrān* (Pahlavi *awērān*) 'ruined, destroyed'— a meaning given by Dehkhoda for some examples."

4. On an excellent formulation of precarity/precariousness, on who counts more and who less, who deserves mourning and what deaths are "acceptable," see Judith Butler's (2010) *Frames of War* (i.e., on the calculus of American vs. Iraqi lives in the representation of war and death).

5. And while maintaining skepticism over the colonial narrative here, still the idea is relevant: that the worse things are, the less indebted tenants can repay their (imposed) debts and rents, and, in a vicious cycle, the worse things get for them. It is thus debt and not famine that ultimately begets death here.

6. Note the parallel to Marx's (1888) famous construction of workers as people with nothing to sell but their labor, people who "live only so long as they find work, and who find work only so long as their labour increases capital. These laborers, who must sell themselves piecemeal, are a commodity, like every other article of commerce, and are consequently exposed to all the vicissitudes of competition, to all the fluctuations of the market." Note further, however, that the people who form the core focus of *this* book, as I've underscored, have not even proper labor, as situated within formalized structures of production, to sell—they are neither citizens, all the way, nor legitimate proletarians (nor still children, all the way): hence the sale of the body itself for sex, the collection and sale of other people's garbage, the collection of alms from other people's labor wages, the gleaning of funds from structures outside production, like shrines, stealing, trinket sales, and other efforts within "the informal economy" (following Portes, Castells, and Benton 1989). I will delve into this notion much more deeply later.

7. For context, the 24 Parganas was the same area fled as a child by the afore-

mentioned Masud, the now-adult, polio-stricken parking attendant at Delhi's Hanuman Mandir (who claimed he had "ended his family's poverty" by returning after many years of hard work on the street).

8. These particular records were obtained from the Bangalore-based NGO SATHI, and they documented station arrivals in Delhi from Bihar in a survey from May 2010. The analysis was carried out at the University of Vermont by Brendan Hennessy, supervised by me.

9. Of course a "district" is a rather arbitrary category—as, in many cases, may be a "village"—and our reliance on districts and other such units is an incidental or accidental fact about colonial and precolonial administration and the national state (see as usual B. Anderson's [1991]) chapter "Census, Map, Museum"); it may be an area defined on the basis of some other criterion— say a particular language group, or surrounding a certain geographical feature (a floodplain, a plateau), or perhaps revolving around some older administrative criterion, like a principality—that is in fact the active territorial domain we should be looking at. It is also undeniably the case, however, that, a district *is* a category based on which people make decisions and take actions—a *reified* construct, and thus socially real in a way that has material entailments.

10. As I assert elsewhere, though trafficking is inseparable from the "running away" phenomenon, in these pages I deal primarily with children who have left home of their own accord, and not at the hands of traffickers. Nonetheless, in many areas, the bulk of children distant from home have been led away, or bought, by traffickers for labor.

11. In more than one site (and by more than one NGO), the *draw* of NGO services in the big cities was cited as a *reason* for running away.

12. Mahali is an Austroasiatic language with some thirty-five thousand speakers that may occupy an intermediate position between Santali and Mundari and features properties of both.

13. And *lands*. Running away to India from Nepal and Bangladesh is common. Running away from Pakistan is extremely rare, though I met at least one child who claimed a lowland Pakistan origin. Among the children, interestingly, *Pakistan* was a kind of taunt, suggesting "boondock" or "bumpkin" identity, and *Bengali* was used as a blanket term to describe a wide variety of languages contrasting to Hindi.

14. And an even wider net, beyond the runaway question, could be cast in the consideration of "crisis," in grappling with what counts as "collapse"—take the dead children of Dharmahsati Gandaman (Mashrakh region, Saran District, Bihar). A simple accident, one might say, that the cooking oil in their school lunch was replaced by pesticide, or an anomalous, random act of malice. They

were no more likely to die than any other children, right? Or wrong? What if the ingestion of monocrotophos organophosphates really is more, what if indeed it is a dark and apt metaphor for the ubiquity of industrial agriculture, its poisons, of the Green Revolution itself in the body's most intimate reaches? In such a light, it is perhaps no coincidence at all that these pesticide poisonings, which saw children staggering home to die in their parents' arms, happened in one of the runaway-densest districts of Bihar in the runaway-densest state of them all. Agrarian globalization literally killed the children. The massive spate of farmer suicides in Maharashtra is another testament to a type of rural crisis in India, and despite the raging and inconclusive debate over the role of debt induced by Monsanto's genetically modified rice, what sort of crisis it testifies to is not clear. What is clear is that people want to talk about it in those terms, and that matters in its own right. Toxic water, massacre, repeated gang rape, abduction, corruption, continuing slavery, indiscriminate killing: what is clear is that such implications of historical change for individual lives can be at the very least understood to be the hints, signs, and traces of a large and troubling experience of rural anomie.

Chapter 4. Death and the Urchin

I am indebted to colleagues at the Society for Cultural Anthropology's 2012 conference and to my writing group—including Kabir Tambar, Kelda Jamison, Danilyn Rutherford, Ben Eastman and Vicki Brennan—for their extensive comments on earlier iterations of this work. Some portions of a previous version of this chapter have appeared in a different form in *Ethnos* 80, no. 2 (2014) and in a paper presented at the American Ethnological Society, also 2012.

1. Train No. 2392 to Nalanda District, around 1:15 p.m.
2. Though stories circulated among the station children at the time that, in the preceding year, Petu's father had visited the station more than once to try to dissuade him from staying in Delhi, or to persuade him to return.
3. The children or their bereaved may frame what I am calling *social* contact, however, as spectral or phantasmic, as in visions or possessions, or apparitions.
4. For the sake of ethics, to prevent any breach of private identity, I confirmed with lawyers in India that this is a public document.
5. In some countries, like Haiti (see Kovats-Bernat 2006), Brazil (see Hecht 1998; and Scheper-Hughes 2005), and Egypt, street children are purged from public view by death squads that summarily execute them.

6. As a rule, in line with ethical convention, I change the names of all subjects in my study. In Petu's case, however, given the public circulation of information about his death in High Court proceedings and the press, I have used the actual nickname and name. I follow suit with the other deaths discussed here, but never with living children, nor with any others whose identities are not publicly iterated.

7. Both *bhut-voot* and *haath-vaath sar-var* are examples of a type of reduplication in Hindi-Urdu, "partial reduplication" (see Kachru 2006), that implies the presence of a thing accompanied by its cognate forms, i.e., x and "all that."

8. A community that was known as Dom, possibly related, if obliquely, is widely accepted to have been the caste category or group in South Asia whose name the European and West Asian community that would be known as Rom/Roma, or "Gypsies," brought with them.

9. See Parry's (1994) *Death in Banaras* for a now-classic account of this space and Huberman (2012) for a recent ethnography of working children at the ghats.

Chapter 5. Runaway Train

1. And the transformation and passage that the train enacts is twofold: not only have they arrived in heaven/Agra, but where before they were seven or eight, they are now, as they shake off the dust from their fall, suddenly pubescent boys of thirteen or fourteen. Thus the train provides passage not only from nowhere to somewhere representing place in its most compelling materiality but also from childhood to something that comes after it. Certainly such images affect and effect life in complicated ways; how they do so is not in the purview of this chapter, but it is taken up by Aguiar (2011) and Kerr (2003).

2. One of the banyan members of the Ficus family (*F. religiosa*) associated with Vishnu, Lakshmi, and Hanuman, and with various other sacred meanings, the pipal is also the "sacred fig" and the "bo tree" under one of which the Buddha received enlightenment; I thus interpret the fact that it is such a central axis (mundi?), even in the stations (which have fully-elaborated sacred geographies as well, mosques and temples and shrines built into the platforms), as rooted in the larger ritual complex in which it is accorded value, and as one of the elements that makes the station's space culturally legible as self-bounded and not disconnected from rites of movement-as-pilgrimage.

3. It is also of note that the activity they are engaging in here, a form of garbage collection and public cleaning, is one that was until recently and in the eyes of many passenger observers from village and city alike the domain of pro-

fessional, caste-based scavengers or sweepers (i.e., Bhangis or Balmikis), and that this reading is grafted onto the perception of the status that such children should be accorded (despite the trans-caste composition of their groups).

4. See Prakash (1990) and Torabully and Carter (2002) on coolies and bonded labor in and from India.

5. Marx himself, of course, emphasized the role of the railways in the development of capital, even in tracts as short as "History as Class Struggle" (1848), and even in India itself; in the *New York Daily Tribune* in 1853, in a piece entitled "The Future Results of British Rule in India," he offered that "modern industry, resulting from the railway system, will dissolve the hereditary divisions of labor upon which rest the Indian castes, those decisive impediments to Indian progress and Indian power" (as quoted in Bear 2007: 2).

6. As quoted in Vinay Lal's introduction to the 1995 reprint (Naidu [1915] 1995).

Chapter 6. Remand to Rehabilitation

1. I take *beggary,* which is used throughout the texts I analyze here, to refer to a whole bundle of attributes, behaviors, and appearances—the assemblage that constitutes *being* a beggar and looking like one, rather than the *act* of begging alone.

2. A shrine caretaker, however, equated pickpockets with faqirs.

3. A cleaning both literal and figurative, as a vendor explains: "People from the Aga Khan Foundation sweep the area all day. They have also given dustbins to all shopkeepers here to keep the streets clean. So the Aga Khan Foundation is doing good work here." A *khadim,* however, said, "There was an order from the Supreme Court that the area will be cleaned but nothing happened. We were happy with the order because we want this place to be cleaned. That order had nothing to do with Aga Khan. Aga Khan is a recent entry."

4. This is not to suggest that the Chishti order is itself bound or limited to South Asia but rather that the Chishti interaction with Delhi's urban space is a unique one. Indeed, Ernst and Lawrence (2002), in their authoritative account of the Chishtiyya, note that "the features of the Sufi master presented in Indo-Muslim biographical literature could be replicated elsewhere; what is distinctly Indian about them is elusive, especially since the earliest literature is self-consciously Muslim and only by circumstance Indian" (70). Elsewhere they suggest that the Chishtiyya form "the major brotherhood to be identified solely with the subcontinent," though it is "neither exclusively Indian nor Islamic" (8).

5. While this is not the space either to define or to trace the history of Sufism, it should be noted that Islamic societies have generated an extraordinarily well-

elaborated set of mystical movements (of which the Chishtiyya are but one example), many or most of which, despite their division into fraternal orders (*tariqat*) and chains of spiritual succession (*silsilah*), from the Naqshbandi to the Qadiriyya, have strong historical ties to each other and substantial impacts on non-Sufistic practice more widely. See (for example) Schimmel 1975 for a basic background.

6. Humayun (b. 1508, d. 1556, reigned 1530–1555) was the second Mughal emperor. His tomb, considered an architectural treasure, is widely considered the antecedent and model for the Taj Mahal and is a UNESCO World Heritage Site. The Nizamuddin renewal is framed as part of the larger process that began with its restoration by the Aga Khan Foundation and local and national partners over many decades.

7. One of the vendors of such cards was markedly hostile when approached with questions for this chapter.

8. And still a site of contention: a shopkeeper mentioned an "issue with Aga Khan Foundation. They wanted to remove the pedestrian shopkeepers from the wall outside Ghalib's Mazar. But these are old sellers. Where would they go? They said give us an alternative space and we will leave. But later the Aga Khan Foundation did not take any action and let them be there. I know about the plans of the Aga Khan Foundation for cleaning this place and I have also seen their map."

9. See Werbner 2003; Marsden 2005; Bowen 1993; and particularly Kugle 2007 for more on the phenomenology of Muslim/Sufi experience, corporality, and identity.

10. A charbagh is a garden in four quadrants imbued with a polyvalent set of cosmological meanings widely discussed and debated in Muslim treatises and which is sometimes meant or said to be evocative of the gardens of paradise.

11. It is of note that woman *are* permitted in the actual tomb in the Dargah Sharif of Nizamuddin's predecessor in Ajmer.

12. Views on these khadims (shrine functionaries) are deeply ambivalent and contingent. A female beggar in Nizamuddin suggested they keep much of the money intended for the poor for themselves, among other imputations.

13. Though these elements, as Ernst and Lawrence (2002) suggest, cannot be deemed "Hindu" *or* "non-Islamic," such is the syncretism they evince. Indeed, they are part of a local cultural complex that is formed not along religious lines but by virtue of distinctive traditions rooted in place and praxis.

14. A faqir—not to be confused with the ethnic group of the same name or for the generalized meaning of "beggar"—is a Muslim ascetic mendicant mystic, often itinerant and usually voluntarily poor, donations to whom can bestow

blessings (*barakat*) on donors. Muslim and Hindu traditions of mendicancy in South Asia have long been dynamically intertwined; each has interacted with and influenced the other to the point that Hindu mendicants frequent Muslim poles of pilgrimage, and the other way around. Nonetheless, there are critical differences, although boundaries are murky: it might be generalized that the question of attachment is an axis of difference; in Hindu traditions and practices, a guru is a conduit to truth but still (in theory) a piece of the world's evanescence, and, even where she/he attains "godhead" status as a unique and sufficient object of devotion (Kabir, Mirabai, Shirdi Sai Baba), she/he is usually still and doctrinally one incarnation among many possible, whereas in Sufi Muslim mendicancies, attachment to a *pir* (spiritual teacher) by a *murid* (adept) is fundamental, the pir is historically unique, and the pir may even be framed as having some kind of cosmic permanence, as a feature of the universe, a *qutb*. If the pir is part of a chain (*isnad* or silsilah) of any kind it is usually one of narrative or spiritual succession. Isma'ilism features a kind of middle ground between these: the imam is both historical and ahistorical; he is always embodied uniquely but a manifestation of a permanent structure of the universe; there is always an imam as long as there is time. Among Hindu mendicants, renunciation of worldly attachment is central. Among Sufi mendicants or faqirs, desire is often to be channeled into devotion for a given saint—and ecstatic practice—but not necessarily negated. Devotion to the saint can be described in terms that are nearly or explicitly erotic (Kugle 2007). Such ecstatic practice may also contrast in a limited sense with Hindu traditions of stillness, silence, and isolation. In both traditions, annihilation or dissolution of some sort (*fana'a* in Sufism and *samadhi* in Hinduism) is central.

15. I say "in the ideal" in part because this is about a set of shared *views* and discourses of such spaces, but pilgrimage sites may sometimes in reality be more pointedly intense zones of addiction and child sexual exploitation, as the NGOs Equations and ECPAT point out in *Unholy Nexus: Male Child Sexual Exploitation in Pilgrim Tourism Sites in India* (2008). This report focuses only on Hindu pilgrimage zones in south central India.

16. The *Church Missionary Intelligencer,* vol. 6, in an entry by Henrietta Neele (1881), gives a very thorough account of the history and establishment of the orphanage at Agarpara, or "Augurpara," occasioned by severe floods in Bengal (1832–1833) and famine in the Northwest Provinces of British India (1834). "It would be impossible," writes Neele, "for one of her missionary spirit to dwell in the gross heathen darkness without doing something to shed abroad the light of the gospel. In a letter of hers of a later date she writes: 'What a field for exertion is this awfully dark pagan land!'" (742). The mission held a boys'

school, a girls' orphanage, and a church and spearheaded vigorous campaigns at conversion through a long chain of successors to Mrs. Wilson.

17. The diary goes on with some interesting observations about the preindependence moment:

> Poona being an orthodox stronghold of the Mahasahba, said Mrs. D., is a difficult place for the welfare-worker. . . . As regards the general feeling of the people in her district, she said that there was still friendship among many of the poorer people, but that she was beginning to notice a certain mocking which had not existed before. The impression was that there was (likely to be) [inserted with carat] some trouble brewing in this interim period between war and peace. There was a sense of stagnation and corruption everywhere . . .
>
> She advised me to travel about India as quietly and unobtrusively as possible, as I should then be more likely to inspire trust and confidence. India, she said, distrusted and was sick of officialdom and all its trappings. Simplicity and humility were more likely to evoke a response these days . . .
>
> N.B. The army seems much more in touch with events than does officialdom.

Chapter 7. Concluding Thoughts, Final Words, and Big Pictures

1. It is notable that the place called Banta and the district called Tata Nagar are now as they were already then in the state of Jharkhand, generated in part by "Adivasi" and "tribal" politics in 2000.
2. *Jhuggi* denotes a thatch hut, tent, or temporary dwelling, generally inhabited by poor and itinerant classes. The child has also taken on the name of the temple that characterizes his village.
3. Tuck (2009) proposes an alternative, however, that might synthesize and address all this, in the form of "desire-centered research."
4. And indeed on some level I feel, as does Nordstrom (2007), that it is an affront to the project of ethnography and to the subjects who suffer to suggest that such a thing might be by design.
5. It is of note that Scott (1985) too, as well as Ortner (1995), warns against "romanticizing" resistance (29).
6. Even child-focused sex tourism could be framed that way, if starting from the proposition that there is a unitary set of "first-world" consumers which decrees that the labor (i.e., the child sex workers) is itself also in some conceptual sense unitary. Of course, the flow of capital is not at all unitary nor as organized in

such a circumstance as it could be with electronic waste work. Child domestic labor is a better analog to sex work in this sense. And of course, sometimes child domestic labor is sex work *and* domestic work.

7. He uses the term *bekār,* for which "useless" is perhaps a better—but less apt—translation.

References

Abadía-Barrero, César E. "Growing Up in a World with AIDS: Social Advantages of Having AIDS in Brazil." *AIDS Care* 14, no. 3 (2002): 417–23.

Abu-Lughod, Lila. "The Romance of Resistance: Tracing Transformations of Power through Bedouin Women." *American Ethnologist* 17, no. 1 (1990): 41–55.

Adiga, Aravind. *The White Tiger*. New York: Free Press, 2008.

Adorno, Theodor W. *Negative Dialectics*. Translated by E. B. Ashton. New York: Continuum, 1973.

Adorno, Theodor W. "Cultural Criticism and Society." In *Prisms*, edited by Samuel and Shierry Weber, 17–34. Cambridge: MIT Press, 1982.

Adorno, Theodor W. *Aesthetic Theory*. Translated by C. Lenhardt. New York: Routledge and Keegan Paul, 1984.

Agamben, Giorgio. *Homo Sacer: Sovereign Power and Bare Life*. Palo Alto, CA: Stanford University Press, 1998.

Agee, James, and Walker Evans. *Let Us Now Praise Famous Men*. Boston: Houghton-Mifflin, 1941.

Agha, Asif. *Language and Social Relations*. Cambridge: Cambridge University Press, 2007.

Aguiar, Marian. *Tracking Modernity: India's Railway and the Culture of Mobility*. Minneapolis: University of Minnesota Press, 2011.

Ahearn, Laura. "Agency." *Journal of Linguistic Anthropology* 9, no. 1–2 (1999): 12–15.

Anderson, Benedict. *Imagined Communities: Reflections on the Origin and Spread of Nationalism*. London: Verso, 1991.

Anderson, Clare. *Legible Bodies: Race, Criminality and Colonialism in South Asia*. Oxford: Berg, 2004.

Andrew, Sir William. *Indian Railways: As Connected with the British Empire in the East*. London: W. H. Allen, 1884.

Appadurai, Arjun. "Spectral Housing and Urban Cleansing: Notes on Millennial Mumbai." *Public Culture* 12, no. 3 (2000): 626–51.

Aptekar, Lewis. *Street Children of Cali*. Durham, NC: Duke University Press, 1988.

Ariès, Philippe. *Centuries of Childhood: A Social History of Family Life*. New York: Vintage Books, 1962.

Augé, Marc. *In the Metro*. Minneapolis: University of Minnesota Press, 2002.

Baker, Rachel, and Catherine Panter-Brick. "A Comparative Perspective on Children's 'Careers' and Abandonment in Nepal." In *Abandoned Children*, edited by Catherine Panter-Brick and Malcolm T. Smith, 161–81. Cambridge: Cambridge University Press, 2000.

Bakhtin, M. M. "Epic and Novel." In *The Dialogic Imagination*, edited by Michael Holquist, 3–40. Austin: University of Texas Press, 1941.

Bakhtin, M. M. *Speech Genres and Other Late Essays*. Translated by Vern W. McGee. Edited by Caryl Emerson and Michael Holquist. Austin: University of Texas Press, 1986.

Balagopalan, Sarada. *Inhabiting "Childhood": Children, Labour, and Schooling in Postcolonial India*. London: Palgrave, 2014.

Bardhan, Pranab. *Land, Labor, and Rural Poverty: Essays in Development Economics*. New York: Columbia University Press, 1984.

Bardhan, Pranab. *Poverty, Agrarian Structure, and Political Economy in India*. Oxford: Oxford University Press, 2003.

Bateson, Gregory. *Naven: A Survey of the Problems Suggested by a Composite Picture of the Culture of a New Guinea Tribe Drawn from Three Points of View*. Cambridge: Cambridge University Press, 1936.

Bauman, Richard. *A World of Others' Words: Cross-Cultural Perspectives on Intertextuality*. Oxford: Blackwell, 2004.

Baviskar, Amita. "Between Violence and Desire: Space, Power, and Identity in the Making of Metropolitan Delhi." *International Social Science Journal* 175 (2003): 89–98.

Baviskar, Amita. *In the Belly of the River: Tribal Conflicts Over Development in the Narmada Valley*. Oxford: Oxford University Press, 2004.

Bear, Laura. *Lines of the Nation: Indian Railway Workers, Bureaucracy, and the Intimate Historical Self*. New York: Columbia University Press, 2007.

Behar, Ruth. *The Vulnerable Observer: Anthropology That Breaks Your Heart*. Boston: Beacon Press, 1996.

Berman, Edward H. *The Ideology of Philanthropy: The Influence of the Carnegie, Ford and Rockefeller Foundations on American Foreign Policy*. Albany: State University of New York Press, 1983.

Bernard, H. Russell. *Research Methods in Anthropology: Qualitative and Quantitative Approaches*. 2nd ed. Walnut Creek: AltaMira, 1994.

Bhattacharya, Bela. *Slums and Pavement Dwellers of Calcutta Metropolis*. Calcutta: Aparna Book Distributors, 1996.

Biehl, João. *Vita: Life in a Zone of Social Abandonment.* Berkeley: University of California Press, 2005.

Bloch, Maurice, and Jonathan Parry, eds. *Death and the Regeneration of Life.* Cambridge: Cambridge University Press, 1982.

Blom Hansen, Thomas. *The Saffron Wave: Democracy and Hindu Nationalism in Modern India.* Princeton: Princeton University Press, 1999.

Bohannan, Laura [as Elenore Smith Bowen]. *Return to Laughter: An Anthropological Novel.* New York, Anchor, 1954.

Boissevain, Jeremy. *Friends of Friends: Networks, Manipulators, and Coalitions.* Oxford: Blackwell, 1974.

Boltanski, Luc. *Distant Suffering: Morality, Media and Politics.* Cambridge: Cambridge University Press, 1999.

Bornstein, Erica. *Disquieting Gifts: Humanitarianism in New Delhi.* Stanford, CA: Stanford University Press, 2012.

Bose, Pablo S. "Living the Way the World Does: Global Indians and the Reshaping of Kolkata," *Annals of the Association of American Geographers* 104, no. 2 (2014): 391–400.

Bose, Pablo S. *Urban Development in India: Global Indians in the Remaking of Kolkata.* London and New York: Routledge, 2015.

Bose, Sugata. *South Asia and World Capitalism.* Oxford: Oxford University Press, 1991.

Bose, Sugata. *Peasant Labor and Colonial Capital.* The New Cambridge History of India Series. Cambridge: Cambridge University Press, 1993.

Bose, Sugata, ed. *Credit, Markets and the Agrarian Economy of Colonial India.* Delhi: Oxford University Press, 1994.

Bourdieu, Pierre. *Outline of a Theory of Practice.* Cambridge Studies in Social and Cultural Anthropology. Cambridge: Cambridge University Press, 1977 (French ed., 1972).

Bourdieu, Pierre. "The Forms of Capital." In *Handbook of Theory and Research for the Sociology of Education,* edited by John Richardson, 241–58. New York: Greenwood Press, 1986.

Bourdillon, M. F. C. *Poor, Harrassed, but Very Much Alive: An Account of Street People and Their Organisation.* Gweru, Zimbabwe: Mambo Press, 1991.

Bourgois, Philippe. "Confronting Anthropology, Education, and Inner-City Apartheid." *American Anthropologist* 98, no. 2 (1996): 249–65.

Bourgois, Philippe. "Families and Children in Pain in the U.S. Inner City." In *Small Wars: The Cultural Politics of Childhood,* edited by Nancy Scheper-Hughes and Carolyn Sargent, 331–51. Berkeley: University of California Press, 1998.

Bowen, John Richard. *Muslims through Discourse: Religion and Ritual in Gayo Society.* Princeton, NJ: Princeton University Press, 1993.

Boyden, Jo, and Judith Ennew. *Children in Focus: A Manual for Participatory Research.* Stockholm: Rädda Barnen, 1997.

Braudel, Fernand. "History and the Social Sciences: The Longue Durée. Translated by Immanuel Wallerstein. *Review* 32, no. 2 (2009): 171–203.

Breitbart, M. "Participatory Research." In *Key Methods in Geography*, edited by Nicholas Clifford and Gill Valentine, 161–78. London: Sage, 2003.

Breman, Jan. *Control of Land and Labour in Colonial Java.* Dordrecht: Foris, 1983.

Breman, Jan. "The Informal Sector." In *The Oxford India Companion to Sociology and Anthropology*, edited by Veena Das, Andre Beteille, and T. N. Madan, 1287–1318. Oxford: Oxford University Press, 2003.

Bucholtz, Mary. "Youth and Cultural Practice." *Annual Review of Anthropology* 31 (2002): 525–52.

Butler, Judith. *Frames of War: When Is Life Grievable?* London: Verso, 2010.

Calcutta Christian Observer. Calcutta: Printed at the Baptist Mission Press, 1837.

Campos, Regina, Marcela Raffaelli, and Walter Ude. "Social Networks and Daily Activities of Street Youth in Belo Horizonte, Brazil." *Child Development* 65, no. 2 (2008): 319–30.

Carsten, Janet. *After Kinship.* Cambridge: Cambridge University Press, 2004.

Celan, Paul. "Todesfuge." In *Der Sand und Den Urnen.* Vienna: Sexl, 1948.

Chakrabarty, Dipesh. "Open Space/Public Place: Garbage, Modernity and India." *South Asia* 14, no. 1 (1991): 15–32.

Chakrabarty, Dipesh. "Of Garbage, Modernity and the Citizen's Gaze." *Economic and Political Weekly* 27, no. 10/11 (1992): 541–47.

Chakraborty, Dipanjan. "Bari Theke Paliye (Runaway from home)." *Dipanjan's Random Muses.* August 12, 2006. http://dipanjanc.blogspot.com/2006/08 /bari-theke-paliye-runaway-from-home.html.

Chakraborty, Shibram. *Bari Theke Paliye.* Calcutta: Book Emporium, 1946.

Chamberlain, Gethin. "Delhi Sweeps Streets of Beggars as India Prepares for Commonwealth Games." *Observer.* November 7, 2009. Accessed February 24, 2017. https://www.theguardian.com/world/2009/nov/08/delhi-common wealth-games-beggars-police.

Chatterjee, Partha. *Nationalist Thought and the Colonial World: A Derivative Discourse.* London: Zed, 1986.

Chatterjee, Partha. *The Nation and Its Fragments: Colonial and Postcolonial Histories.* Princeton, NJ: Princeton University Press, 1993.

Chauhan, Brij Raj. "Village Community." In *The Oxford India Companion to Sociology and Anthropology*, edited by Veena Das, Andre Beteille, and T. N. Madan, 409–57. Oxford: Oxford University Press, 2003.

Chitralekha. *Ordinary People, Extraordinary Violence.* Oxfordshire: Routledge, 2012.

Christian, David. *Maps of Time: An Introduction to Big History.* Berkeley: University of California Press, 2004.

Çinar, Alev, and Thomas Bender. *Urban Imaginaries: Locating the Modern City.* Minneapolis: University of Minnesota Press, 2007.

Clifford and Marcus. *Writing Culture: The Poetics and Politics of Ethnography.* Berkeley: University of California Press, 1986.

Cohen, David William, and E. S. Atieno Odhambo. "Silences of the Living, Orations of the Dead: The Struggle in Kenya for S. M. Otieno's Body, 20 December 1986 to 23 May 1987." In *Between History and Histories: The Making of Silences and Commemorations,* edited by Gerald Sider and Gavin Smith, 180–98. Toronto: University of Toronto Press, 1997.

Comaroff, Jean. *Body of Power, Spirit of Resistance: The Culture and History of a South African People.* Chicago: University of Chicago Press, 1985.

Comaroff, Jean, and John L. Comaroff. *Ethnography and the Historical Imagination.* New York: Routledge, 1992.

Comaroff, Jean, and John Comaroff. *Law and Disorder in the Postcolony.* Chicago: University of Chicago Press, 2006.

Conklin, Lynn, and Beth Morgan. "Babies, Bodies, and the Production of Personhood in North America and a Native Amazonian Society." *Ethos* 24, no. 4 (1996): 657–94.

Conticini, Alessandro. "Surfing in the Air: A Grounded Theory of the Dynamics of Street Life and Its Policy Implications." *Journal of International Development* 20, no. 4 (2008): 413–36.

Conticini, Alessandro, and David Hulme. "Escaping Violence, Seeking Freedom: Why Children in Bangladesh Migrate to the Street." *Development and Change* 38, no. 2 (2007): 201–27.

Crapanzano, Vincent. *Tuhami: Portrait of a Moroccan.* Chicago: University of Chicago Press, 1980.

Cresswell, Tim. *On the Move: Mobility in the Modern World.* New York: Routledge, 2006.

Cullather, Nick. *The Hungry World: America's Cold War Battle against Poverty in Asia.* Cambridge, MA: Harvard University Press, 2010.

Cunningham, Hugh. *The Children of the Poor: Representations of Childhood since the Seventeenth Century.* Cambridge, MA: Blackwell, 1991.

Cunningham, Hugh. *Children and Childhood in Western Society since 1500.* London: Longman, 1995.

Cunningham, Hugh. *The Invention of Childhood.* London: BBC Books, 2006.

Daniel, E. Valentine. *Charred Lullabies: Chapters in an Anthropography of Violence.* Princeton: Princeton University Press, 1996.

Das, Veena. "Our Work to Cry, Your Work to Listen." In *Mirrors of Violence: Communities, Riots and Survivors in South Asia,* edited by Veena Das. Delhi: Oxford University Press, 1990.

Das, Veena. *Life and Words.* Berkeley: University of California Press, 2006.

Davis, Mike. *Late Victorian Holocausts.* London: Verso, 2001.

De Certeau, Michel. *The Practice of Everyday Life.* Berkeley: University of California Press, 1984.

Deleuze, Giles, and Félix Guattari. *A Thousand Plateaus: Capitalism and Schizophrenia.* Minneapolis: University of Minnesota Press, 1987.

Dhaul, Laxmi, Pallee, and Anoop Kamath. *The Dargah of Nizamuddin Auliya.* New Delhi: Rupa, 2006.

Dickens, Charles. "The Signal-Man, in Mugby Junction." *All the Year Round,* Special Christmas Edition (December 25, 1866): 20–25.

Dirks, Nicholas B. *Castes of Mind: Colonialism and the Making of Modern India.* Princeton, NJ: Princeton University Press, 2001.

Dos Santos, Benedito Rodrigues. *Ungovernable Children: Runaways, Homeless Youths, and Street Children in New York and São Paolo.* PhD diss., University of California at Berkeley, 2002.

Douglas, Mary. *Purity and Danger: An Analysis of the Concepts of Pollution and Taboo.* London: Routledge and Kegan Paul, 1966.

Driskell, D. *Creating Better Cities with Children and Youth: A Manual for Participation.* London: United Nations Educational, Scientific, and Cultural Organization/Earthscan, 2002.

Durkheim, Émile. *The Elementary Forms of the Religious Life: A Study in Religious Sociology.* Translated by Joseph Ward Swain. London: G. Allen and Unwin, 1926.

Durkheim, Émile, and George Simpson. *On the Division of Labor in Society.* New York: Macmillan, 1933.

Eco, Umberto. "Towards a Semiotic Inquiry into the Television Message." *Working Papers in Cultural Studies* 3 (Autumn 1972): 103–21.

Ennew Judith. *Street and Working Children: A Guide to Planning.* London: Save the Children Fund, 1994.

EQUATIONS/ECPAT International. *Unholy Nexus: Male Child Sexual Exploitation in Pilgrim Tourism Sites in India.* Bangalore and Bangkok, 2008.

Ernst, Carl W., and Bruce B. Lawrence. *Sufi Martyrs of Love: The Chishti Order in South Asia and Beyond.* New York: Palgrave Macmillan, 2002.

Fadiman, Anne. *The Spirit Catches You and You Fall Down.* New York: Farrar, Straus, and Giroux, 1997.

Faleiro, Sonia. "Why Do So Many Indian Children Go Missing?" *New York Times*, November 29, 2017.

Farmer, Paul. "On Suffering and Structural Violence: The View from Below." *Daedalus* 124, no. 1 (1996): 261–83.

Farmer, Paul. *Infections and Inequalities: The Modern Plagues*. Berkeley: University of California Press, 2001.

Farmer, Paul. "An Anthropology of Structural Violence." *Current Anthropology* 45, no. 3 (June 2004): 305–25.

Farmer, Paul. "The Second Life of Sickness: On Structural Violence and Cultural Humility." *Human Organization* 75, no. 4 (2016): 279–88.

Fegan, Brian. 1986. "Tenants' Non-Violent Resistance to Landowner Claims in a Central Luzon Village." *Journal of Peasant Studies* 13, no. 2 (1986): 87–106.

Ferguson, James. *Give a Man a Fish: Reflections on the New Politics of Distribution*. Durham, NC: Duke University Press, 2015.

Finkelstein, Marni. *With No Direction Home: Homeless Youth on the Road and in the Streets*. Case Studies on Contemporary Social Issues. Belmont, CA: Thomson Wadsworth, 2005.

Foucault, Michel. *Discipline and Punish: The Birth of the Prison*. New York: Random House, 1975.

Foucault, Michel. *The History of Sexuality*. Vol. 1. Paris: Editions Gallimard, 1978.

Foucault, Michel. *Power/Knowledge: Selected Interviews and Other Writings, 1972–1977*. New York: Pantheon Books, 1980.

Foucault, Michel. "The Subject and Power." *Critical Inquiry* 8 (1982): 777–95.

Foucault, Michel. *The History of Sexuality, Vol. 3: The Care of the Self*. New York: Vintage, 1988.

Foucault, Michel. *Discipline and Punish: The Birth of the Prison*. 1977. Repr., New York: Vintage, 1995.

Foucault, Michel. *Abnormal: Lectures at the Collège de France, 1974–1975*. New York: Picador, 2004.

Freed, Ruth, and Stanley Freed. "Ghost Illness of Children in North India." *Medical Anthropology* 12, no. 4 (1990): 401–17.

Freed, Ruth, and Stanley Freed. *Ghosts: Life and Death in North India*. New York: American Museum of Natural History, 1993.

Gaborieau, Marc. "Criticizing the Sufis: The Debate in Early-Nineteenth-Century India." In *Islamic Mysticism Contested: Thirteen Centuries of Controversies and Polemics,* edited by Frederick de Jong and Bernd Radtke, 452–67. Leiden: Brill, 1999.

Ganguly-Scrase, Ruchira. "Victims and Agents: Young People's Understanding of Their Social World in an Urban Neighbourhood in India." *Young* 15 (2007): 321–41.

Geertz, Clifford. *Works and Lives: The Anthropologist as Author.* Palo Alto, CA: Stanford University Press, 1988.

Ghertner, Asher D. *Rule by Aesthetics: World-class City Making in Delhi.* Oxford: Oxford University Press, 2015.

Gigengack, Roy. "Critical Omissions: How Street Children Studies Can Address Self-Destructive Agency." In *Research with Children,* edited by Pia Monrad Christensen and Allison James, 205-19. London: Routledge, 2008.

Gillespie, Susan D. "Personhood, Agency, and Mortuary Ritual: A Case Study from the Ancient Maya." *Journal of Anthropological Archaeology* 20 (2001): 73-112.

Glauser, Benno. "Street Children: Deconstructing a Construct." In *Constructing and Reconstructing Childhood: Contemporary Issues in the Sociological Study of Childhood,* edited by Allison James and Alan Prout, 145-64. London: Falmer Press, 1997.

Goldstein, Daniel M. *The Spectacular City: Violence and Performance in Urban Bolivia.* Durham, NC: Duke University Press, 2004.

Goldstein, Donna M. "Nothing Bad Intended: Child Discipline, Punishment, and Survival in a Shantytown in Rio de Janeiro, Brazil." In *Small Wars: The Cultural Politics of Childhood,* edited by Nancy Scheper-Hughes and Carolyn Sargent, 389-415. Berkeley: University of California Press, 1998.

Gordon, Avery F. *Ghostly Matters: Haunting and the Sociological Imagination.* Minneapolis: University of Minnesota Press, 1997.

Gough, Katherine, and Monica Franch. "Spaces of the Street: Socio-spatial Mobility and Exclusion of Youth in Recife." *Children's Geographies* 3, no. 3 (2005): 149-66.

Graeber, David. *Debt: The First 5000 Years.* New York: Melville House Printing, 2011.

Gramsci, Antonio. *Selections from the Prison Notebooks.* Edited by Quintin Hoare. 1934; repr., London: Lawrence and Wishart, 2005.

Greenblatt, Stephen. *Cultural Mobility: A Manifesto.* Cambridge: Cambridge University Press, 2010.

Grew, Raymond. "On Seeking Global History's Inner Child." *Journal of Social History* 38 (2005): 849-58.

Guha, Ranajit. *Elementary Aspects of Peasant Insurgency in Colonial India.* Delhi: Oxford, 1983.

Gupta, Narayani. "The Indian City." In *The Oxford India Companion to Sociology and Anthropology,* edited by Veena Das, Andre Beteille, and T. N. Madan, 458-76. Oxford: Oxford University Press, 2003.

Hardt, Michael, and Antonio Negri. *Empire.* Cambridge, MA: Harvard University Press, 2000.

Harris, Gardiner. "Message from a Sadhu: Detach from Family to Avoid Sadness." *New York Times,* February 11, 2013.

Hart, R. A. *Children's Participation.* London: United Nations Children's Fund/ Earthscan, 1992.

Harvey, David. *The Urban Experience.* Baltimore: Johns Hopkins University Press, 1985.

Harvey, David. *The Condition of Postmodernity: An Enquiry into the Origins of Social Change.* London: Blackwell, 1989.

Hausner, Sandra. *Wandering with Sadhus: Ascetics of the Hindu Himalayas.* Bloomington: Indiana University Press, 2007.

Hearn, Gordon R. *The Seven Cities of Delhi.* London: W. Thacker, 1906.

Hebdige, Dick. *Subculture: The Meaning of Style.* London: Routledge, 1979.

Hecht, Tobias. *At Home in the Street: Street Children of Northeast Brazil.* Cambridge: Cambridge University Press, 1998.

Hecht, Tobias. "In Search of Brazil's Street Children." In *Abandoned Children,* edited by Catherine Panter-Brick and Malcolm T. Smith, 146–60. Cambridge: Cambridge University Press, 2000.

Hertz, Robert. *Death and the Right Hand.* 1907; repr., Glencoe, IL: Free Press, 1960.

Heywood, Colin. *A History of Childhood.* London: Polity, 2002.

Hobbes, Robert George. "Scenes in the Cities and Wilds of Hindostan." 1852. Handwritten manuscript, British Library, India Office Archive.

Höjdestrand, Tova. *Needed by Nobody: Homelessness and Humanness in Post-Socialist Russia.* Ithaca, NY: Cornell University Press, 2009.

Huberman, Jenny. "Working and Playing Banaras: A Study of Tourist Encounters, Sentimental Journeys, and the Business of Visitation." PhD diss., University of Chicago, 2006.

Huberman, Jenny. *Ambivalent Encounters: Childhood, Tourism, and Social Change in Banaras, India.* Trenton, NJ: Rutgers University Press, 2012.

Hull, Matthew. "The File: Agency, Authority, and Autography in a Pakistan Bureaucracy." *Language and Communication* 23, no. 3/4 (2003): 287–314.

Hull, Matthew. *Government of Paper: The Materiality of Bureaucracy in Urban Pakistan.* Berkeley: University of California Press, 2012.

Human Rights Watch. *Police Abuse and Killings of Street Children in India.* New York: Human Rights Watch, 1996.

India Imperial Legislative Council. *Abstract of the Proceedings of the Council of the Governor-General of India, Assembled for the Purpose of Making Laws and Regulations.* Calcutta: Legislative Council, 1875.

India Imperial Legislative Council. *Abstract of the Proceedings of the Council of the*

Governor-General of India, Assembled for the Purpose of Making Laws and Regulations. Office of the Superintendent of Government Print, 1890.

Irfan, August 25, 2011. *Mail Today.*

Jakobson, Roman. "The Poetry of Grammar and the Grammar of Poetry." In *Language in Literature,* edited by K. Pomorska and S. Rudy, 121-44. Cambridge, MA: Belknap, 1987.

James, Allison, and Chris Jenks. "Public Perceptions of Childhood Criminality." *British Journal of Sociology* 47, no. 2 (1996): 315-31.

Jenks, Chris. *Childhood.* 2nd ed. London: Routledge, 2005.

Jodhka, Surinder. "Agrarian Structures and Their Transformations." In *The Oxford India Companion to Sociology and Anthropology,* edited by Veena Das, Andre Beteille, and T. N. Madan, 1213-1242. Oxford: Oxford University Press, 2003.

Johnstone, Barbara. *Stories, Community and Place: Narratives from Middle America.* Bloomington: Indiana University Press, 1990.

Jones, Gareth A., Elsa Herrera, and Sarah Thomas De Benitez. "Tears, Trauma and Suicide: Everyday Violence among Street Youth in Puebla, Mexico." *Bulletin of Latin American Research* 26, no. 4 (2007): 462-79.

Kachru, Braj. *The Handbook of World Englishes.* New York: Wiley, 2006.

Kara, Siddharth. *Bonded Labor: Tackling the System of Slavery in South Asia.* New York: Columbia University Press, 2012.

Karlekar, Malavika. "Domestic Violence." In *The Oxford India Companion to Sociology and Anthropology,* edited by Veena Das, Andre Beteille, and T. N. Madan, 1127-57. Oxford: Oxford University Press, 2003.

Kelly, Tobias. "A Life Less Miserable?" *HAU: Journal of Ethnographic Theory* 3, no. 1 (2013): 213-16.

Kerr, Ian. *Building the Railways of the Raj.* Delhi: Oxford, 1995.

Kerr, Ian. "Representation and Representations of the Railways of Colonial and Post-colonial South Asia." *Modern Asian Studies* 37 (2003): 287-326.

Kerr, Ian. *Engines of Change: The Railroads That Made India.* Westport, CT: Praeger, 2007.

Kilbride, Philip, Enos Njeru, and Collete Suda. *Voices of Children in Search of a Childhood.* Westport, CT: Bergin and Garvey, 2000.

King, Roger, and Gavin Kendall. *The State, Democracy and Globalization.* Basingstoke: Palgrave Macmillan, 2004.

Korbin, Jill E. "Children, Childhoods, and Violence." *Annual Review of Anthropology* 32 (2003): 431-46.

Kovats-Bernat, J. Christopher. *Sleeping Rough in Port-au-Prince: An Ethnography of Street Children and Violence in Haiti.* Gainesville: University Press of Florida, 2006.

Kracke, W. H. Kagwahiv. "Mourning II: Ghosts, Grief, and Reminiscences." *Ethos* 16, no. 2 (1988): 209–22.

Kroeber, A. L. "Disposal of the Dead." *American Anthropologist* 29 (1927): 308–15.

Kugle, Scott A. *Sufis and Saints' Bodies: Mysticism, Corporeality, and Sacred Power in Islam.* Chapel Hill: University of North Carolina Press, 2007.

Kumarappa, J. M. *Our Beggar Problem.* Bombay: Padma Publications, 1945.

Lancy, David. *The Anthropology of Childhood: Cherubs, Chattel, Changelings.* Cambridge: Cambridge University Press, 2008.

Larymore, A. D. *Report of the Alipore Reformatory School for the Year 1880.* Calcutta: Bengal Secretariat Press: 1881.

Lawrence, Bruce. *Nizam Ad-Din Awliya: Morals for the Heart.* Mahwah, NJ: Paulist Press, 1991.

LeFebvre, Henri. *La production de l'espace.* Paris: Editions Anthropos, 1974.

Legg, Stephen. *Spaces of Colonialism: Delhi's Urban Governmentalities.* Oxford: Blackwell, 2007.

Levi, Anthony. *Renaissance and Reformation: The Intellectual Genesis.* New Haven: Yale University Press, 2002.

Levine, Nancy A. "Alternative Kinship, Marriage, and Reproduction." *Annual Review of Anthropology* 37 (2008): 375–89.

Lewnes, Alexia. *Misplaced: New York City's Street Children.* New York: Xenium Press, 2001.

Li, Tania Murray. "Indigeneity, Capitalism, and the Management of Dispossession." *Current Anthropology* 51, no. 3 (June 2010): 385–414.

Low, Setha M. "The Anthropology of Cities: Imagining and Theorizing the City." *Annual Review of Anthropology* 26 (1996): 383–409.

Lucchini, R. "The Street and Its Image." *Childhood* 3, no. 2 (1996a): 235–46.

Lucchini, R. "Theory, Method and Triangulation in the Study of Street Children." *Childhood* 3, no. 2 (1996b): 167–70.

Margolin, Gayla, and Elana B. Gordon. "The Effects of Family and Community Violence on Children." *Annual Review of Psychology* 51 (2000): 445–79.

Marsden, Magnus. *Living Islam: Muslim Religious Experience in Pakistan's North-West Frontier.* New York: Cambridge University Press, 2005.

Marushiakova, Elena, and Vesselin Popov. *Gypsies in Central Asia and the Caucasus.* Basingstoke, UK: Palgrave Macmillan, 2016.

Marx, Karl. "Eighteenth Brumaire of Louis Napoleon." In *Die Revolution,* edited by Joseph Weydemeyer. New York, 1852.

Marx, Karl. "The Future Results of British Rule in India." *New York Daily Tribune,* August 8, 1853.

Marx, Karl. *Capital: A Critique of Political Economy.* Vol. 1. Moscow: Progress Publishers, 1867.

Marx, Karl, and Frederick Engels. *Manifesto of the Communist Party.* Translated by Frederick Engels. Chicago: Charles H. Kerr and Company, 1888.

Massey, Doreen. *Space, Place, and Gender.* Oxford: Polity Press, 1994.

Masud, Muhammad Khalid. "The Growth and Development of the Tablighi Jama 'at in India." In *Travelers in Faith: Studies of the Tablighi Jama 'at as a Transnational Islamic Movement for Faith Renewal.* Leiden: Brill, 2000: 3–43.

McFadyen, Lori. *Voices from the Street: An Ethnography of Indian Street Children.* Delhi: Hope India Publications, 2004.

Menon, Dilip. "The Blindness of Insight: Why Communalism in India Is about Caste." In *Indian Political Thought,* edited by Aakash Singha and Silika Mohapatra, 123–36. New York: Routledge, 2010.

Menon-Sen, Kalyani, and Gautam Bhan. *Swept Off the Map: Surviving Eviction and Resettlement in Delhi.* New Delhi: Yoda Press, 2008.

Metcalf, Barbara Daly. "Living Hadith in the Tablighi Jama'at." *Journal of Asian Studies* 52, no. 3 (1993): 584–608.

Metcalf, Peter, and Richard Huntington. *Celebrations of Death: The Anthropology of Mortuary Ritual.* Cambridge: Cambridge University Press, 1991.

Mickelson, Roslyn Arlin, ed. *Children on the Streets of the Americas: Globalization, Homelessness and Education in the United States, Brazil and Cuba.* London: Routledge, 2000.

Mines, Diane P. "Hindu Periods of Death 'Impurity.'" In *India through Hindu Categories,* edited by McKim Marriott, 103–30. New Delhi: Sage, 1990.

Mintz, Steven. *Huck's Raft: A History of American Childhood.* Cambridge, MA: Harvard University Press, 2004.

Mol, Annemarie. *Body Multiple: Ontology in Medical Practice.* Durham, NC: Duke University Press, 2002.

Moncrieffe, Joy. "When Labels Stigmatize: Encounters with 'Street Children' and 'Restavecs' in Haiti." In *The Power of Labelling: How People Are Categorized and Why It Matters,* edited by Joy Moncrieffe and Rosalind Eyben, 80–96. London: Earthscan, 2007.

Montgomery, Heather. "Abandonment and Child Prostitution in a Thai Slum Community." In *Abandoned Children,* edited by Catherine Panter-Brick and Malcolm T. Smith, 182–98. Cambridge: Cambridge University Press, 2000.

Morrow, Virginia. "Using Qualitative Methods to Elicit Young People's Perspectives on Their Environments: Some Ideas for Community Health Initiatives." *Health Education Research* 16, no. 3 (2001): 255–68.

Murugan, Perumal. *Seasons of the Palm.* Chennai: Tara Books, 2004.

Nagy, M. "The Child's Theories Concerning Death." *Pedagogical Seminary and Journal of Genetic Psychology* 73 (1948): 3–27.

Naidu, M. Pauparao. *The History of Railway Thieves, with Illustrations and Hints on Detection.* 4th ed. Edited by Vinay Lal. 1915; repr., Gurgaon: Vintage Press, 1995.

Nandy, Ashis. *The Intimate Enemy: Loss and Recovery of Self Under Colonialism.* Delhi: Oxford University Press, 1983.

Neele, Henrietta. *The Church Missionary Intelligencer.* Vol. 6. London: Church Missionary House, 1881.

Nieuwenhuys, Olga. *Children's Lifeworlds: Gender, Welfare and Labour in the Developing World.* London: Routledge, 1994.

Nieuwenhuys, Olga. "The Domestic Economy and the Exploitation of Children's Work: The Case of Kerala." *International Journal of Children's Rights* 3, no. 2 (1995): 213–25.

Nieuwenhuys, Olga. "The Paradox of Child Labor and Anthropology." *Annual Review of Anthropology* 25 (1996): 237–51.

Nordstrom, Carolyn. *Shadows of War: Violence, Power, and International Profiteering in the Twenty-First Century.* Berkeley: University of California Press, 2004.

Nordstrom, Carolyn. *Global Outlaws: Crime, Money, and Power in the Contemporary World.* Berkeley: University of California Press, 2007.

Nordstrom, Carolyn. "The Bard." In *Anthropology Off the Shelf: Anthropologists on Writing,* ed. Alisse Waterston and Maria D. Vesperi, 35–45. Malden, MA: Wiley-Blackwell, 2009.

Ochs, Elinor, and Lisa Capps. *Living Narrative: Creating Lives in Everyday Storytelling.* Cambridge, MA: Harvard University Press, 2001.

O'Malley, L. S. S. *The Bengal District Gazetteer for Muzaffarpur.* Calcutta: Bengal Secretariat Book Depot, 1905.

Ong, Aihwa. *Flexible Citizenship: The Cultural Logics of Transnationality.* Durham, NC: Duke University Press, 1999.

Ortner, Sherry. "Resistance and the Problem of Ethnographic Refusal." *Comparative Studies in Society and History* 37, no. 1 (1995): 173–93.

Ortner, Sherry. "Dark Anthropology and Its Others." *HAU: Journal of Ethnographic Theory* 6, no. 1 (2016): 47–73.

Pain, Rachel. "Social Geography: Participatory Research." *Progress in Human Geography* 28, no. 5 (2004): 652–63.

Pande, Rajendra. *Street Children of India: A Situational Analysis.* Allahabad, India: Chugh Publications, 1991.

Pandian, Anand. *Crooked Stalks: Cultivating Virtue in South India.* Durham, NC: Duke University Press, 2009.

Panter-Brick, Catherine, and Malcolm T. Smith, eds. *Abandoned Children.* Cambridge: Cambridge University Press, 2000.

Panter-Brick, Catherine. "Street Children, Human Rights, and Public Health: A

Critique and Future Directions." *Annual Review of Anthropology* 31 (2002): 147–71.

Parry, Jonathan P. *Death in Banaras.* Cambridge: Cambridge University Press, 1994.

Patel, Sheela. "Street Children, Hotel Boys and Children of Pavement Dwellers and Construction Workers in Bombay: How They Meet Their Daily Needs." *Environment and Urbanization* 2, no. 2 (1990): 9–26.

Pellegrin, Paolo. "As I Was Dying." https://www.magnumphotos.com/newsroom /conflict/paolo-pellegrin-as-i-was-dying/.

Phillips, W. S. K. *Street Children in India.* Jaipur: Rawat Publications, 1994.

Pierson, Arthur T. *The Missionary Review of the World.* New York: Funk and Wagnalls, 1897.

Pinch, William R. *Warrior Ascetics and Indian Empires.* Cambridge: Cambridge University Press, 2006.

Platts, John T. *A Dictionary of Urdu, Classical Hindi, and English.* London: W. H. Allen, 1884.

Portes, Alejandro, and Manuel Castells. "World Underneath: The Origins, Dynamics, and Effects of the Informal Economy." In *The Informal Economy,* edited by Alejandro Portes, Manuel Castells, and Lauren A. Benton, 11–37. Baltimore: Johns Hopkins University Press, 1989.

Portes, Alejandro, Manuel Castells, and Lauren A. Benton, eds. *The Informal Economy.* Baltimore: Johns Hopkins University Press, 1989.

Posner, Marc. "Hungry Hearts: Runaway and Homeless Youth in the United States." In *Children on the Streets of the Americas: Globalization, Homelessness and Education in the United States, Brazil and Cuba,* edited by Roslyn Arlin Mickelson, 247–56. London: Routledge, 2000.

Pouchepadass, Jacques. *Champaran and Gandhi: Planters, Peasants, and Gandhian Politics.* Oxford: Oxford University Press, 1999.

Povinelli, Elizabeth. *The Cunning of Recognition: Indigenous Alterities and the Making of Multicultural Australia.* Durham, NC: Duke University Press, 2002.

Praharaj, Ira. "Indigo Plantation in India Under the British Rule." 2012. https:// www.youtube.com/watch?v=8AqWlH74B74.

Prakash, Gyan. *Bonded Histories: Genealogies of Labor Servitude in India.* Cambridge: Cambridge University Press, 1990.

Presner, Todd. *Mobile Modernity: Germans, Jews, Trains.* New York: Columbia, 2007.

Quesada, James. "Suffering Child: An Embodiment of War and Its Aftermath in Post-Sandinista Nicaragua." *Medical Anthropology Quarterly* 12, no. 1 (1998): 51–73.

Rabinow, Paul. *Reflections on Fieldwork in Morocco.* Berkeley and Los Angeles: University of California Press, 1977.

Radhakrishna, Meena. *Dishonoured by History: "Criminal Tribes" and British Colonial Policy.* Delhi: Orient Longman, 2001.

Raffaelli, Marcela. "The Family Situation of Street Youth in Latin America: A Cross-National Review." *International Social Work* 40, no. 1 (1997): 89–100.

Raffaelli, Marcela, et al. "Sexual Practices and Attitudes of Street Youth in Belo Horizonte, Brazil." *Social Science Medicine* 37, no. 5 (1993): 661–70.

Raffaelli, Marcela, et al. "HIV-Related Knowledge and Risk Behaviors of Street Youth in Belo Horizonte, Brazil." *AIDS Education and Prevention* 7, no. 4 (1995): 287–97.

Rao, Ursula. "Making the Global City: Urban Citizenship at the Margins of Delhi." *Ethos* 75, no. 4 (2010): 402–24.

Rawls, John. *A Theory of Justice.* Cambridge, MA: Harvard University Press, 1971.

Ray, Rabindra. *Naxalites and Their Ideology.* 3rd. ed. Oxford: Oxford University Press, 2012.

Report on the Administration of Bengal, 1902–03. Calcutta: Bengal Secretariat Book Depot, 1904.

Ribeiro, Moneda Oliveira, and Maria Helena Trench Ciampone. "Homeless Children: The Lives of a Group of Brazilian Street Children." *Journal of Advanced Nursing* 35, no. 1 (2001): 42–49.

Robbins, Joel. "Beyond the Suffering Subject: Toward an Anthropology of the Good." *Journal of the Royal Anthropological Institute* 19, no. 3 (2013): 447–62.

Rosaldo, Renato. Culture and Truth: The Remaking of Social Analysis. Boston: Beacon Press, 1989.

Roy, Ananya. *City Requiem, Calcutta: Gender and the Politics of Poverty.* Minneapolis: University of Minnesota Press, 2003.

Roy, Ananya. *Poverty Capital: Microfinance and the Making of Development.* New York: Routledge, 2010.

Roy, Arundhati. *The God of Small Things: A Novel.* New York: Random House, 1997.

Roy, Arundhati. "The NGO-ization of Resistance." 2014. *Massalijn.* http://massa lijn.nl/new/the-ngo-ization-of-resistance/.

Roy, Mallarika S. *Gender and Racial Politics in India: Magical Moments of Naxalbari (1967–1975).* New York: Routledge, 2010.

Sadana, Rashmi. "On the Delhi Metro: An Ethnographic View." *Economic and Political Weekly,* 45, no. 46 (2010): 77–83.

Sainath, Palagummi. *Everybody Loves a Good Drought: Stories from India's Poorest Districts.* Delhi: Penguin, 1996.

Sandel, Michael J. *Justice: What's the Right Thing to Do?* New York: Farrar, Straus, and Giroux, 2009.

Sanyal, Kalyan. *Rethinking Capitalist Development: Primitive Accumulation, Governmentality and Post-colonial Capitalism.* New Delhi: Routledge India, 2007.

Sassen, Saskia. *Expulsions: Brutality and Complexity in the Global Economy.* Cambridge, MA: Harvard University Press, 2014.

Scheper-Hughes, Nancy. *Death without Weeping: The Violence of Everyday Life in Brazil.* Berkeley: University of California Press, 1992.

Scheper-Hughes, Nancy. "Dangerous and Endangered Youth: Social Structures and Determinants of Violence." *Annals of the New York Academy of Science* 1036 (2005): 13–46.

Scheper-Hughes, Nancy, and Daniel Hoffman. "Brazilian Apartheid: Street Kids and the Struggle for Urban Space." In *Small Wars: The Cultural Politics of Childhood,* edited by Nancy Scheper-Hughes and Carolyn Sargent, 352–88. Berkeley: University of California Press: 1998.

Scheper-Hughes, Nancy, and Carolyn Sargent, eds. *Small Wars: The Cultural Politics of Childhood.* Berkeley: University of California Press, 1999.

Schiffrin, Deborah. *Approaches to Discourse.* Oxford: Blackwell, 1994.

Schimmel, Annemarie. *Mystical Dimensions of Islam.* Chapel Hill: University of North Carolina Press, 1975.

Schivelbusch, Wolfgang. *The Railway Journey: The Industrialization of Time and Space in the Nineteenth Century.* Berkeley: University of California Press, 1977.

Scott, James C. *Weapons of the Weak: Everyday Forms of Peasant Resistance.* New Haven: Yale University Press, 1985.

Scott, James C. *Domination and the Arts of Resistance: Hidden Transcripts.* New Haven: Yale University Press, 1990.

Sen, Amartya. *Development as Freedom.* New York: Alfred A. Knopf, 1999.

Sen, Satadru. *Colonial Childhoods: The Juvenile Periphery of India, 1850–1945.* London: Anthem, 2005.

Seth, Rajeev, Atul Kotwal, and K. K. Ganguly. "Street and Working Children of Delhi, India, Misusing Toluene: An Ethnographic Exploration." *Substance Use and Misuse* 40, no. 11 (2005): 1659–80.

Shanahan, Suzanne. "Lost and Found: The Sociological Ambivalence towards Childhood." *Annual Review of Sociology* 33 (2008): 407–28.

Shyrock, Andrew, and Daniel Smail. *Deep History: The Architecture of Past and Present.* Berkeley: University of California Press, 2011.

Sibley, David. "Families and Domestic Routines: Constructing the Boundaries of

Childhood." In *Mapping the Subject: Geographies of Cultural Transformation,* edited by Steve Pile and N. J. Thrift, 123-37. London: Routledge, 1995.

Sivaramakrishnan, Karthik. "Some Intellectual Genealogies for the Concept of Everyday Resistance." *American Anthropologist* 107, no. 3 (2005): 346-55.

Slesnick, Natasha. *Our Runaway and Homeless Youth: A Guide to Understanding.* Westport, CT: Praeger, 2004.

Smart, Alan, and Josephine Smart. "Urbanization and the Global Perspective." *Annual Review of Anthropology* 32 (2003): 263-85.

Spivak, Gayatri Chakravorty. "Can the Subaltern Speak?" In *Marxism and the Interpretation of Culture,* ed. Cary Nelson and Lawrence Grossberg, 271-313. Champaign: University of Illinois Press, 1988.

Srinivas, M. N. *The Remembered Village.* Berkeley: University of California Press, 1976.

Srivastava, Sanjay. *Passionate Modernity: Sexuality, Class and Consumption in India.* New Delhi: Routledge, 2007.

Steinberg, Jonah. *Isma'ili Modern: Globalization and Identity in a Muslim Community.* Chapel Hill: University of North Carolina Press, 2011.

Steinberg, Jonah. "The Social Life of Death on Delhi's Streets: Unclaimed Souls, Pollutive Bodies, Dead Kin and the Kinless Dead." *Ethnos* 80, no. 2 (2013): 248-71.

Stern, Robert W. *Changing India: Bourgeois Revolution on the Subcontinent.* Cambridge: Cambridge University Press, 1993.

Stevenson, Lisa. *Life Beside Itself: Imagining Care in the Canadian Arctic.* Berkeley: University of California Press, 2014.

Stewart, Kathleen. *Ordinary Affects.* Durham, NC: Duke University Press, 2007.

Tamm, M. E., and A. Granqvist. "The Meaning of Death for Children and Adolescents: A Phenomenographic Study of Drawings." *Death Studies* 19, no. 3 (1995): 203-22.

Tandon, Sneh Lata. "Report on the Survey of the Beggars in Delhi." Unpublished manuscript, 2007. University of Delhi, New Delhi.

Tarique, Mohammad. *Modern Indian History.* New Delhi: Tata McGraw Hill, 2008.

Thomas, Nigel, and Claire O'Kane. "The Ethics of Participatory Research with Children." *Children and Society* 12, no. 5 (2006): 336-48.

Thorner, David. *Investment in Empire: British Railway and Steam Shipping Enterprise in India.* Philadelphia: University of Pennsylvania Press, 1950.

Tiwari, Ramaswarup. *Railways in Modern India.* Bombay: New Book, 1941.

Torabully, Khal, and Marina Carter. *Coolitude: An Anthology of the Indian Labour Diaspora.* London: Anthem, 2002.

Tsing, Anna. *Friction: An Ethnography of Global Connection.* Princeton, NJ: Princeton University Press, 2007.

Tuck, Eve. "Suspending Damage: A Letter to Communities." *Harvard Educational Review* 79, no. 3 (2009), 409–27.

Turner, Victor. *The Ritual Process: Structure and Anti-structure.* Piscataway, NJ: AldineTransaction, 1969.

Turner, Victor. *Dramas, Fields, and Metaphors: Symbolic Action in Human Society.* Ithaca, NY: Cornell University Press, 1975.

Turner, Victor. "Death and the Dead in the Pilgrimage Process." In *Blazing the Trail: Way Marks in the Exploration of Symbols,* edited by Edith Turner, 29–47. Flagstaff: University of Arizona Press, 1992.

Uberoi, Patricia L. "The Family in India: Beyond the Nuclear vs. Joint Debate." In *The Oxford India Companion to Sociology and Anthropology,* edited by Veena Das, Andre Beteille, and T. N. Madan, 1061–1103. Oxford: Oxford University Press, 2003.

Urban, Greg. *A Discourse-Centered Approach to Culture: Native South American Myths and Rituals.* Austin: University of Texas Press, 1991.

Urban, Greg. "Entextualization, Replication, and Power." In *Natural Histories of Discourse,* edited by Michael Silverstein and Greg Urban, 21–44. Chicago: University of Chicago Press, 1996.

Urban, Greg. *Metaculture: How Culture Moves through the World.* Minneapolis: University of Minnesota Press, 2001.

Urry, John. *Mobilities.* Cambridge: Polity, 2007.

van der Ploeg, Jan, and Evert Scholte. *Homeless Youth.* Working with Children and Adolescents Series. London: Sage, 1997.

van der Veer, Peter. *Religious Nationalism: Hindus and Muslims in India.* Berkeley: University of California Press, 1994.

Vannini, Phillip, ed. *The Cultures of Alternative Mobilities.* Surrey: Ashgate, 2009.

Veale, Angela, Max Taylor, and Carol Linehan. "Psychological Perspectives of 'Abandoned' and 'Abandoning' Street Children." In *Abandoned Children,* edited by Catherine Panter-Brick and Malcolm T. Smith, 131–45. Cambridge: Cambridge University Press, 2000.

Verdery, Katherine. *The Political Lives of Dead Bodies.* New York: Columbia University Press, 1999.

Verma, Gita Dewan. *Slumming India: A Chronicle of Slums and Their Saviours.* New Delhi: Penguin Books, 2002.

Vigil, James Diego. "Urban Violence and Street Gangs." *Annual Review of Anthropology* 32 (2003): 225–42.

Viswanathan, Gauri. *Masks of Conquest: Literary Study and British Rule in India.* New York: Columbia University Press, 2014.

Vološinov, Valentin. *Marxism and the Philosophy of Language.* Cambridge, MA: Harvard University Press, 1973 [1929].

Voluntary Health Association of India. *Seen, but Not Heard: India's Marginalised, Neglected, and Vulnerable Children.* New Delhi: VHAI, 2002.

Walker, Harry, and Iza Kavedžija. "Values of happiness." *HAU: Journal of Ethnographic Theory* 5, no. 3 (2015), 1–23.

Wallerstein, Immanuel. *World Systems Analysis: An Introduction.* Durham, NC: Duke University Press, 2004.

Watson, James L. "Of Flesh and Bones: The Management of Death Pollution in Cantonese Society." In *Death and the Regeneration of Life,* edited by Maurice Bloch and Jonathan Parry, 155–86. Cambridge: Cambridge University Press, 1982.

Weiner, Myron. "Migration." In *The Oxford India Companion to Sociology and Anthropology,* edited by Veena Das, Andre Beteille, and T. N. Madan, 262–82. Oxford: Oxford University Press, 2003.

Werbner, Pnina. *Pilgrims of Love: The Anthropology of a Global Sufi Cult.* Indiana University Press, 2003.

White, Christine Pelzer. "Everyday Resistance, Socialist Revolution and Rural Development: The Vietnamese Case." *Journal of Peasant Studies* 13, no. 2 (1986): 49–63.

Wiesel, Elie. *Night.* New York: Hill and Wang, 1960.

Wilkerson, Isabelle. *The Warmth of Other Suns: The Epic Story of America's Great Migration.* New York: Random House, 2010.

Wilkins, William Joseph. *Hindu Mythology.* London: Thacker, Spink, 1882.

Wilkinson, Iain, and Arthur Kleinman. *A Passion for Society: How We Think about Human Suffering.* Oakland: University of California Press, 2016.

Worthman, Carol M., and Catherine Panter-Brick. "Homeless Street Children in Nepal: Use of Allostatic Load to Assess the Burden of Childhood Adversity." *Development and Psychopathology* 20, no. 1 (2009): 233–55.

Index